Facing Climate Change

Building South Africa's Strategy

Stefan Raubenheimer

2011

For my mother, for her help, interest and encouragement.

This book was made possible by funding from ANSA-Africa
and the World Bank Institute

ISBN 978-1-920409-52-4
First published 2011
Editing by Glenda Younge
Design and cover design by Jacana Media
Production by Idasa Media

Bound and printed by Unity Press, Cape Town

Contents

Acknowledgements

The Long-Term Mitigation Scenarios (LTMS), a South African government led process, had far-reaching impact both in South Africa and abroad. Its unique approach to process, combining stakeholder involvement and research, is central to its success. This also meant that the project was the product of a large team effort. There were many people who gave a large amount of their time and effort to make the project a success.[1]

I acknowledge the team at the Department of Environmental Affairs and Tourism who led the LTMS work, in particular Joanne Yawitch, Peter Lukey, Tsietsi Mahema, Kelebogile Moroka and many others. They managed the process and provided leadership and insight in measure.

The government asked the Energy Research Centre (ERC) of the University of Cape Town to both manage the process, and to lead the research-driven component. Harald Winkler took on the task of overall leadership of the LTMS project, and he and the ERC team deserve credit for this effort. The research outputs, which were co-ordinated by the ERC through four teams made up of our best researchers nationally, produced cutting edge data, which will be useful for years to come.

Lastly, the LTMS was a stakeholder process. The stakeholder management was the responsibility of Tokiso, the organisation which provided the event management and three facilitators: Edwin Mohlalehi, Pascal Moloi

[1] The Long-Term Mitigation Scenario process was a South African government initiative, which was run by the Department of Environmental Affairs and Tourism (as it then was). It was sanctioned by the Cabinet of the South African government and funded internally. All intellectual property in the products of the LTMS remains the property of the South African government.

and me. Our task was to manage the unfolding stakeholder process, which in large part is the subject of this account. The many individuals from government, the private sector, labour and civil society who took part in the process as members of the Scenario Building Team (SBT) gave time and attention without reward, and are recognised in this account for their contribution. In the high-level process that followed, Chief Executive Officers (CEOs) and leaders from government departments, civil society organisations and the major labour federations, applied their minds to the outputs of the SBT: the combined result was a remarkable emerging consensus.

In the process of developing this account, a number of participants from the SBT of the LTMS agreed to add their inputs, and we conducted a series of interviews. These inputs are included verbatim in text boxes throughout the book. I wish to thank the following contributors, who gave time and enthusiasm freely and added nuggets of wisdom to this work. They are (in no particular order):

Laurraine Lotter: CEO, Chemical and Allied Industry Association

Bob Scholes of the Council for Scientific and Industrial Research (CSIR)

Imraan Patel of the Department of Science and Technology

Kilebogile Maroka, then of the Department of Environmental Affairs

Richard Worthington, then of Earthlife Africa

Fred Goede and Herman van der Walt of Sasol

Mandy Rambharos of Eskom

Kevin Nassiep of Central Energy Fund (CEF)

Shaun Vorster, erstwhile Special Adviser to the Minister of Environment and Tourism

Guy Midgley, director of the South African National Botanical Society

I wish to thank the Institute for Democracy in Africa (IDASA), the sponsor of this book and Richard Calland who showed enthusiastic interest and tolerance to delay. I would also like to thank Saliem Fakir most warmly for his engagement in this project, and Kim Coetzee and my mother for their hard work on the text.

This account is presented as a companion piece to Harald Winkler's book, *Taking Action on Climate Change* (University of Cape Town Press, 2009), which presents the technical outputs of the LTMS. In that book all the data and detailed results of the process are gathered and hence will not be repeated in this account.

I echo the sentiments of Harald Winkler in his acknowledgements: 'The approval of the Scenario Document (SBT, 2007) based on that intensive process stands out for me as one of the remarkable products of facilitated process, rigorous research – and the willingness of thinkers from different perspectives to listen to one another and to find a way forward. If everyone from large emitters through government officials to NGOs can continue to work together in this way, there is yet hope that effective action on Climate Change may soon be implemented in South Africa.'

My final word of thanks must go to Harald: our collaboration on the LTMS project was special, memorable and cemented a friendship.

Foreword

The Long-Term Mitigation Scenarios (LTMS), a government-led South African process, approached the mitigation challenge 'from the bottom-up'. It involved a broad range of stakeholders who collectively bolted together both best-practice research and human imagination. The position the South African government took, based on the LTMS, was much hailed, both nationally and internationally. The LTMS is now constantly mentioned, and has had a remarkably lasting impact.

This report explores the *process design* of the LTMS, its implementation and unfolding story, and the lessons learnt. I look at the LMTS as an experiment in building policy through broad consensus-making. I also see the LTMS approach as a way to shape governance that is built on a foundation of social accountability.

I intend this book for a general audience. If you are interested in the challenges that a high-emitting, developing country faces when confronting the almighty task of managing its development path whilst at the same time reducing its greenhouse gas emissions, this book will, in part at least, show possible ways to approach the problem. I hope too that process facilitators, policy-makers and leaders in general find this account useful in other contexts or countries. Because process professionals often don't know much about the technical issues around Climate Change I have not assumed that readers are highly qualified in this area. On the other hand, I apologise to those who are experts, and acknowledge that my own technical knowledge is not always up to the task.

I recommend that this book be read as a companion piece to Harald Winkler's comprehensive book on the technical outcomes of the LTMS, *Taking Action on Climate Change* (UCT Press, 2009)[2]. which is aimed at those qualified and interested in the complex technical issues with which the research teams grappled.

In South Africa, through the LTMS, we achieved a remarkable consensus. This book is the story of how we got there.

---

2 Winkler, H. 2009. *Taking Action on Climate Change*, Cape Town: University of Cape Town Press.

Acronyms

AKST	Agricultural Knowledge, Science and Technology
AR	Assessment Report
BAU	Business as Usual
BUSA	Business Unity South Africa
CAIA	Chemical and Allied Industrial Association
CDP	Current Development Plan
CDT	Current Development Trend
Cosatu	Congress of South African Trade Unions
CSIR	Council for Scientific and Industrial Research
DEAT	Department of Environmental Affairs and Tourism
DPE	Department of Public Enterprises
EIUG	Energy Intensive Users' Group
ERC	Energy Research Centre of the University of Cape Town
Eskom	Electricity Supply Commission
GDP	Gross Domestic Product
GHG	Greenhouse gas
GWC	Growth Without Constraints
IAASTD	The International Assessment of Agriculture Science and Technology for Development
IDC	Inter-Departmental Committee
IMC	Inter-Ministerial Committee
IPCC	Intergovernmental Panel on Climate Change
LADA	The Land Degradation Assessment of Drylands
LCGP	Low Carbon Growth Plan
LTAS	Long-Term Adaptation Scenarios
LTMS	The Long-Term Mitigation Scenarios
Nactu	National Council of Trade Unions
NCCC	National Climate Change Committee
PBMR	Pebble Bed Modular Reactor
PMT	Project Management Team
RBS	Required by Science
SACAN	South African Climate Action Network
SBT	Scenario Building Team
UNFCCC	UN Framework Convention on Climate Change

Chapter 1

Beginnings

> 'There is no greater asset for humanity than the long-term health and well-being of our planet. There can be no goal more crucial to our survival than the protection and nurturing of our natural environment. One of our most urgent challenges as the global community is to convince all nations to join and support the international effort to reduce the emissions of greenhouse gases. I have no doubt that the next few years will be crucial to move us out of an approach of stalling, of avoidance, and of excuses to one where we all accept our responsibility to deal with Climate Change within an inclusive multilateral international framework. Climate Change is a global scourge and requires a unified global partnership for action.'
>
> *Marthinus van Schalkwyk, Minister of Environment and Tourism, April 2005*

When the officials of the Department of Environmental Affairs and Tourism called Harald Winkler and I to Pretoria, we had no idea that the next two years of our working lives would be dominated by the project we had helped to dream up six months earlier. This was in 2006, a red-letter year for Climate Change, with news reports bringing the issue centre stage, world wide. In South Africa at the time, only a handful of people were working on Climate Change.

Leaders in government were in the room when we first met to plan the Long-Term Mitigation Scenarios (LTMS). The Cabinet had, crucially, approved the idea of an exploration of our country's mitigation options, and the Energy Research Centre (ERC) of the University of Cape Town was tasked to lead the study. The ERC was given a remarkable mandate: involve the leading thinkers in South Africa in a participatory research exercise to reveal a set of options for greenhouse gas mitigation over a long-term trajectory. This was government saying: get together the best team you can and show us some options – which sounded simple enough.

This mandate must be seen in the context of a South Africa which was emerging from the first decade of democracy, and gathering speed as a developing economy. In other words, our study was to take place in an atmosphere in which 'growth' was seen by most as untouchable; growth was seen as the panacea to the woes of poverty, and the only game in town was the growth model. South Africa's growth is inextricably linked to a fossil fuel-based energy supply and most captains of industry at the time firmly held the view that fossil-based growth was the only economically viable alternative for the country. So too did most government leaders. The proponents of growth held that we should not commit to reducing our greenhouse gas emissions, as this was really the responsibility of those industrialised countries which emitted significantly more. They held that we needed 'carbon space' to grow, develop and eliminate poverty.

Government officials, our Minister and his adviser, and delegates who had been involved in the international climate negotiations, knew that South Africa was a large emitter relative to its size, and that it could not avoid its responsibility for mitigating its emissions forever.

A small group of increasingly outspoken activists distrusted leadership in government and industry, and criticised the fossil-intensive approach to our development. Some civil society leaders believed that South Africa was a significant polluter and that it should reduce its emissions drastically. They singled out Eskom and Sasol as prime culprits, which resulted in regular clashes in the media.

While some held that nuclear energy was the only option for a growing energy crisis, others opposed any plan that included nuclear energy. Some said that we had to rely on seemingly endless supplies of coal. Yet others were beginning to explore growth decoupled from fossil emissions – a new clean, green South Africa. In the field of energy and emissions, we had a true democracy of opinions.

There was a big problem, however: we had little or no real data. Eskom had done some modelling and thinking about their own greenhouse gas emissions, as had others, but no-one had investigated an integrated countrywide plan.

In this atmosphere of divided opinion, the government started to make some significant noises about its responsibility, as can be seen from the statement of Minister van Schalkwyk at the start of the chapter. Government was, in a way, speaking out 'ahead of the prevailing interest'. It also knew, however, that we needed to learn a great deal more about ourselves before we could plan the actual discharge of that responsibility.

Our study would therefore take place in a country firmly divided on energy issues. We were embarking on a rather high-risk project, given that the many people who were to take part were known for their strong opinions and extensive expertise. When you put a group of people in a room, and get them to talk about Climate Change, energy and development, you can expect the equivalent of a Latino family dinner: strongly expressed opinions and rowdy disagreement, but no resolve – everyone eventually drifting off to bed dimly aware that relationships have been wrecked until the morning at least! The subject of Climate Change, and in particular the thorny question of how to address greenhouse gas emissions, is enormous and complex, and interests, values and differences are writ large. We saw this not only in South Africa, but also on the international stage, where the Climate negotiations were dragging on with no resolution in sight, and with ever-increasing mistrust.

As consultants to the government to deliver the LTMS, Harald and I had proposed a number of ways in which to avoid the pitfalls of process work on a highly divided issue: we would work with Scenarios[3] rather than prescribed plans; we would engage experts rather than representatives; and we would let data tell the story. With a mandate in hand, the study had to be structured to work, to avoid a squabble, and crucially, to produce a set of results. The government knew that South Africa would have to be ready with something to offer at the following climate talks, notably in 2009, when commitments would be under the spotlight.

Harald was to lead the project, and I was to act as lead facilitator of the stakeholder group. As things developed, we collaborated in almost all areas of the project, and in so doing, held the LTMS together. This is the story of the decisions and the processes that unfolded. It seeks to show that the approach itself was key to why the LTMS prevails as a source of data for planning purposes in South Africa, and stands as a model for similar processes elsewhere in the world.

On one level our Scenario approach was aimed at producing options for government, rather than a prescriptive plan. On another level it was aimed at being a mechanism to avoid conflict, and to encourage 'out of the box' thinking. Combining the process with scientific research was the cornerstone of the approach. Equally our stakeholder-driven approach was

[3] I use the word Scenario in its technical sense, as a set of plausible pictures of the future; in the case of the LTMS it developed a different meaning as will be seen later.

designed to get buy-in, and it delivered that, but much more: it turned the LTMS into a lever for action in many sectors, boardrooms and in government itself.

Why did the government of South Africa mandate such a study? What drove the undertaking of such a high-risk approach when it could have commissioned consultants to produce the data? These questions make a brief tangent necessary.

As a starting point an understanding of the Climate Change challenge is important as it shows just how vulnerable South Africa actually is, on a number of levels.

Climate Change and South Africa

The Climate Change challenge is defining our times. It is perhaps the most difficult project facing humanity right now, but it is just one within the broad challenge of sustainability. It is a proxy for so many issues: how we govern ourselves; how we plan our economies; how we care for the next generation; how we produce and consume things; and even our understanding of our very place in the world. Its scale is vast, and it is perhaps even life threatening. The science of Climate Change is immensely complex, and is still emerging. It is also seen as such a threat to fossil fuel industries that they have exploited the uncertainties surrounding certain subsets of Climate Science in order to undermine its central certainties. That said, the world is more or less united on one thing: we do have a problem.

All true solutions to grave problems depend on the will of people to activate solutions that solve the problem in its totality. Actions that amount to a partial solution are pointless. This is why Churchill stated during the World War II: 'It is no use saying, "We are doing our best". You have got to succeed in doing what is necessary.'

The problem is that doing what is necessary is extremely difficult, politically, socially, technologically and economically. So much so that countries are still promising to do only that which they can afford to do, which falls short of what is required. In addition, the time frame for action is terribly tight. The first challenge is political: determining the sequence and proportion of who has to take on the mitigation challenge still stymies the world community. The response so far has been tepid. At the now rather infamous Copenhagen Conference of 2009, there was a widespread understanding of the fact that the stakes were high, yet world leaders backed away from doing what was necessary. The sheer scale of the challenge appears to have frightened us all, and sent nations retreating to positions largely driven by self-interest, domestic politics and mistrust of one other.

2005: a quick tour through Climate Science ... then

The Intergovernmental Panel on Climate Change (IPCC) report of 2001 (Assessment Report 3, or AR3) illustrated, with growing certainty, that human-induced emissions of greenhouse gases were likely to affect the climate. However there was still a significant level of uncertainty on both the predicted impacts, and the level of mitigation (reduction) of greenhouse gas emissions that would be required. The greater certainty achieved in AR4 of 2007 was still in the pipeline.

The cause of global warming and Climate Change is human-induced greenhouse gas emissions, of which there are essentially two components: supply and stock. Greenhouse gases continue to have an impact on the atmosphere for varying periods, hence the stock already in the atmosphere is relevant. For example, CO_2 emitted in Europe during the early period of the industrial revolution still has an impact on the climate today. On the supply side the picture is different, with some developing countries now contributing significantly to emissions.

Science, in the form of AR4, has given us a relatively accurate picture of just how much we have to reduce greenhouse gas emissions, globally, in order to keep us on a (relatively) safe trajectory. Some Climate Change is now unavoidable, and so the question is how much change can we adapt to with safety and hence, what limit should be placed on future emissions to ensure this oucome. Although tolerating a 2° Celsius mean rise (above pre-industrial age levels) in global temperature has its risks, and will certainly cause suffering (especially in the poorer countries), there has been some political agreement that, at the very best, this limit is 'what is necessary' given the greenhouse gasses already emitted. This tentative agreement means, more or less, that by 2050 we have to reduce our global emissions to between 60% and 80% of their 1990 levels.

The period between 2005 and 2008 has been viewed as watershed years for Climate Science. The IPCC stated that the science was now incontrovertible. The review by Sir Nicolas Stern greatly enhanced general understanding, from an economics perspective. But 2005 saw many in industry and politics still adopting a 'wait and see' approach. Awareness of the issue was gaining a ground, and to most leaders, in South Africa and elsewhere, the issue was beginning to appear on various agendas. The work of Stern and the IPCC would motivate and legitimise the LTMS study, but 2005 was early days, and there were still wide differences, even fundamental ones. At many a dinner table the view that this 'climate stuff' was poppycock was still being voiced.

International co-operation

The UN Framework Convention on Climate Change (UNFCCC) of 1992 recognised that most historical emissions, the so-called 'stock' of emissions, were attributable to the developed world. This division between the two worlds, and this differentiation in their respective responsibility, is key to the approach taken in the UN negotiations that followed, and also perhaps the cause of some of the problems. The developing world, quite rightly, looked to the industrial world to act first. It was, after all, their stock that had started the problem. But this bifurcation of our world also meant that the developed world looked on as its emission reductions were eclipsed by the growth in emissions from the developing world.

The major feature of the Kyoto Protocol (signed in 1997) is that it sets binding targets for 37 industrialised countries and the European community for reducing greenhouse gas (GHG) emissions, in line with this 'act first' approach. These targets amount to an average of 5% against 1990 levels over the five-year period 2008–2012. The problem with Kyoto is that the US, the world's biggest supply emitter (until China took the lead in 2009), refused to sign the Protocol. In addition the targets themselves still fell far short of the reductions recommended by science.

The major distinction between the Protocol and the Convention is that while the Convention *encouraged* industrialised countries to stabilise GHG emissions, the Protocol *committed* them to do so. The binding nature of the Protocol was a hard-fought achievement.

Recognising that developed countries are principally responsible for the current high levels of GHG emissions in the atmosphere as a result of more than 150 years of industrial activity, the Protocol placed a heavier burden on developed nations under the principle of 'common but differentiated' responsibilities. So between 1997 and 2012, the end of the first commitment period in the Protocol, no mitigation action was required from the developing world. A second period had to be negotiated before the end of 2009, but most developing countries still excluded themselves from any action for a second period. Copenhagen was intended as the meeting to set the terms for this second period, or to agree on 'Kyoto part 2'. It was therefore something of a crunch-time meeting of the parties and attracted a lot of hype as a result.

In the run up to Copenhagen, and certainly in 2005, there was no talk whatsoever (in South Africa, and elsewhere in the developing world) of taking up mitigation targets from 2012 onwards. All eyes remained fixed on the North to see if they would be first to truly commit to their larger responsibility.

But there was a niggling problem, or rather, a cluster of problems. China especially, was a focus of attention as her supply emissions finally surpassed those of the world's largest emitter, the USA, in 2009. India, too, was developing rapidly and producing rising emissions. There was a growing awareness that if these two countries developed to a 'full Western way of life' for all citizens, we would see perilous Climate Change, simply by virtue of their large populations. At the same time the leaders of China and India maintained the right of their countries to develop. In contrast, mitigation action by these two countries was held up by the US as a precondition for its entry into any mitigation agreement, despite its huge stock emissions. In addition, it was also becoming clear that on its own the developed world was simply not able to act at the pace and scale required by science, and certainly not if the developing countries, which were continually growing, would be emitting 'against' the developed world's efforts. Trade issues and competitiveness, and equal playing fields, were coming into play.

Behind the giants, China and India, sat another clutch of countries: Indonesia, Brazil, Mexico and South Africa – all with growing and significant emissions.

South Africa: important or insignificant?[4]

South Africa was an active participant in the talks and a recognised bridge-builder with a well-regarded negotiations team. But what should its own position on mitigation be? Could it afford to avoid the issue indefinitely?

It is often said that South Africa's emissions are really quite small compared to other countries. On one level this is of course true. However, to get a real picture we need to look at three sets of data: overall or absolute emissions, emissions per capita, and emissions per unit of GDP (emissions intensity). Then we need to compare notes with other countries in the developing country group.

A mere 25 countries produce around 83% of these total 'supply' emissions. It is significant to compare these, which the WRI report does in the table on the next page, published in 2005:[5]

---

4 See Annexure 1 for information on emissions, generally and in South Africa.

5 Ibid.

Top GHG emitting countries (CO_2, CH_4, N_2O, HFC_5, PFC_5, SF_4)		
Country	**$MtCO_2$ equivalent**	**% of world GHGs**
1. United States	6,9328	20.6
2. China	4,938	14.7
3. EU-25	4,725	14.0
4. Russia	1,915	5.7
5. India	1,884	5.6
6. Japan	1,317	3.9
7. Germany	1,009	3.0
8. Brazil	851	2.5
9. Canada	680	2.0
10. United Kingdom	654	1.9
11. Italy	531	1.6
12. South Korea	521	1.5
13. France	513	1.5
14. Mexico	512	1.5
15. Indonesia	503	1.5
16. Australia	491	1.5
17. Ukraine	482	1.4
18. Iran	480	1.4
19. South Africa	417	1.2
20. Spain	381	1.1
21. Poland	381	1.1
22. Turkey	355	1.1
23. Saudi Arabia	341	1.0
24. Argentina	289	0.9
25. Pakistan	285	0.8
Top 25	**27,915**	**83**
Rest of world	**5,751**	**17**
Developed	**17,355**	**52**
Developing	**16,310**	**48**

When non-CO_2 emissions, and more significantly, land use emissions such as those from deforestation are added, some countries such as Indonesia jump up the table. South Africa is not significantly affected. We are up there in the top 20, more or less. But in terms of population and size of economy it is obvious at a glance that we seem to be in the wrong place on this list.

This is explained when one looks at emissions against population (which tells us something about levels of consumption, especially considering that most South Africans emit little due to poverty) and against GDP (which tells us how emissions-efficient production is). South African figures reflect the following:[6]

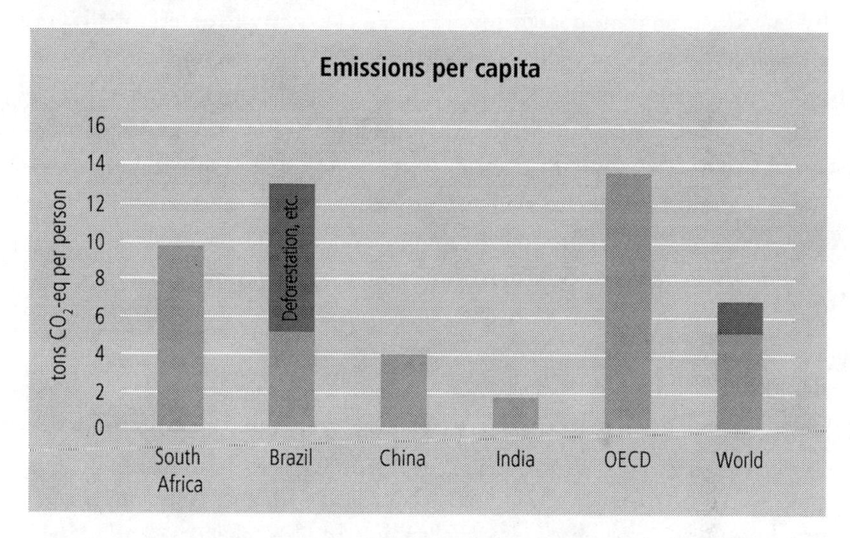

Per capita emissions are well above the world average, and in the case of intensity, South Africa and China are very similar – nastily inefficient.

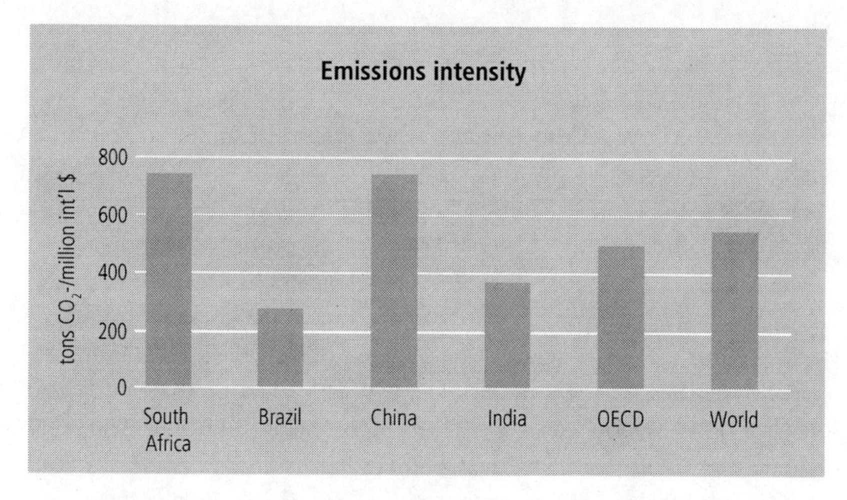

6 LTMS report: can be downloaded at www.erc.uct.ac.za.

It becomes clear that from a comparative perspective, South Africa cannot be ignored as an emitter, and by 2005 the South African government was fully aware of this.

Whilst officially holding the view that significant action was required from the developed world before the developing countries took up obligations, there must certainly have been some concern in South African leadership circles. Our first problem was an economy that was driven largely by burning low-grade coal to supply cheap electricity to an essentially inefficient domestic and industrial consumer base. Our second was a heavy reliance on road transport and liquid fuels, a third of which came from a coal-to-liquids process, which is emissions intensive. As a country we used to have an excess supply of electricity which we sold cheaply to bulk users (such as aluminium smelters) in order to attract foreign and domestic business. Our emissions reflect these economic realities.

For these reasons, South Africa's further development and growth plans seemed coupled to more and more emissions.

Breaking the impasse

Planning emissions reductions is tremendously complex. This activity requires an alignment of investment, policy, action, consumption, monitoring and much more. To plan an economy that aligns with a set emissions trajectory but still stays healthy is very challenging. In 2005 the world was just beginning to develop the tools for this type of planning. Not only did we face a technical challenge, but also one of how to involve people in this technical challenge. Hence the challenge was twofold, that is, technical and process.

Big problems need new solutions: Einstein famously said: 'We can't solve problems by using the same kind of thinking we used when we created them'. I interpret this to mean, in the context of planning such a large task, that both process and action need new ways of thinking. We need new process and technical approaches to crack the impasse in the international negotiations, but we also need to crack the problems at home.

In the South African context at the time, we did have a few problems. A small group in government had been exposed to the international negotiations prior to 2005, and knew and understood the problem. This group, lead by the then Minister of Environment and Tourism, Marthinus van Schalkwyk, his special adviser, and key personnel in his department, had experienced the international negotiations first hand and could see that change was coming. Their position was to prove progressive, forward-looking and prescient. They needed to convince other arms of government that the issue was important. They also needed to engage with industry. Our two largest corporate emitters, Eskom and Sasol, had been sending representatives to the UN conferences, and had started engaging with the issues, but not many others in industry at that time were aware of the climate debates. As for the public at large, the level of awareness was negligible. So it stands to reason that getting a conversation going on mitigation targets or actions was bound to be met with howls of protest, indifference or official shutdown. A much more cautious approach was required.

The development of a comprehensive long-term mitigation plan is, after all, a challenge of social accountability. Given that any mitigation plan will necessarily involve so many levels of society, it should have firm roots with everyone in the 'community', from power players in the economy, to workers and consumers, and from environmentalists to oil barons. This means that a full understanding of long-term, systems-based interest, needs to come from the 'bottom up', and needs to be based on a full understanding of available data, as well as an unflinching embrace of the sheer scale of the challenge. This has to be a national effort; indeed, a national conversation – one in which emotion is stripped out of the equation, and trusted and reliable data inserted in its place. All the agents of change, from the media to politicians, from teachers to Chief Executive Officers (CEOs), would need to be aligned behind the plan.

But how would one start the conversation that would lead to this national consensus?

An idea

It felt to me, at the time, that the *format* of the conversation was all-important. One could simply not succeed in a constructive way by starting with the blunt question of whether or not South Africa should take on a mitigation target, and if so, how ambitious and by when? I felt that we would have to approach the conversation in an emergent way; a way that was in a sense a voyage of discovery for its participants.

I discussed these thoughts with my close colleague Harald Winkler, now Associate Professor at the Energy Research Centre (ERC) of the University of Cape Town. In conversation he introduced the second leg of the problem: the data.

Coincidentally, we lived in the same street in Cape Town, and met in a local coffee shop from time to time to share our thoughts. We had very different backgrounds: Harald is a researcher, and I am a process specialist. Both of us worked in the Climate Change field. Harald had been educating me on the subject of modelling. I listened fascinated as he described how one could enter complex data assumptions into a model (run by a powerful computer) and then do accurate projections into the future. We spoke in particular about an energy/emissions model called MARKAL.[7] In MARKAL one can enter a load of demographic and other inputs (such as costs and performance of technology, through to population and GDP growth assumptions), and in this way 'describe' a country and its economy. Then one can upload present and future cost inputs for the delivery of energy to that economy. Once activated, the model in effect chooses the lowest-cost option on a year-by-year basis, constantly selecting the optimum energy package (seen from a cost point of view) to 'feed' the country and its economy. MARKAL then calculates the cumulating emissions from that energy feed which enables one to see year-by-year growth or decline of emissions right up to the determined horizon (end date) for the model.

I expressed an interest in linking two ideas to the mitigation challenge: one was applying my experience in working with multiple stakeholders in planning processes (I had done this type of work professionally for some years); the second was working with Scenarios as a tool for future studies. I had been reading some of the work of Adam Kahane and Clem Sunter, as well as reports on Louis van der Merwe's work with the South African government. This work was also sometimes done with multiple stakeholders (and hence shared challenges with the work I had been doing), characteristically turning small stakeholder groups into Scenario Building Teams, where future stories were created in order to inform decision-makers in the present. In the South African context the best known of these Scenario processes were the Mont Fleur Scenarios, facilitated before the elections in 1994 by Adam Kahane, and 'built' by some of the people who now run South Africa. These Scenarios tell their own story, and were enormously influential in the transition to post-apartheid South Africa.

Harald and I started connecting the dots. We developed a rough methodology that would enable us to respond to the mandate we would eventually receive to run the LTMS. We had our plan. Sharing our thoughts with our colleagues in government, the thinking rapidly developed along the following lines:

---

7 MARKAL is a generic model tailored by the input data to represent the evolution over a period of usually 40 to 50 years of a specific energy system at the national, regional, provincial or community level.

1. A confrontational and prescriptive format for the 'conversation' would not work, as government would not enjoy being prescribed to, nor would industry players be prepared to start with an assumption that emissions reductions were inevitable. The conversation would have to be speculative, an exploration, open-ended, and conducted in a safe space.
2. At the same time government preferred to engage the players, preferably from all industrial sectors as well as from the social component, and secure both buy-in and capacity building. In government too, other departments would be engaged.
3. It was agreed that the Scenario approach should be used.
4. Independent and non-partisan professionals should facilitate the study.
5. On the knowledge issue, we knew that the available knowledge on mitigation options for South Africa was miniscule. A body of data would have to be built – not a small task. But the study had to be based on accurate science and robust research. Evidence was key to success.
6. Finally it was agreed that the study should look as far as possible into the future.

Context

These early planning conversations, which took place in the opening months of 2006, were really made possible by a number of uniquely South African characteristics.

In the first instance, South Africa has a strong, albeit short, tradition of social accountability in all of its many manifestations. Since our first democratic election in 1994, government has adopted a 'bottom-up' planning approach in most respects, preferring to engage with stakeholders in society to conceive and build policy. Government sees itself as being in partnership with society, and accountable to it, under our new Constitution.

Secondly, the use of Scenario tools has been part of this history, and South African stakeholders have been and still are exposed to Scenario processes, and have used them most effectively. Scenarios were not a new concept.

Thirdly, our government wanted to lead on the Climate Change issue, having been prominent at the Climate Conferences as a bridge-builder. Most importantly, early research had presented our government with a gloomy outlook when it came to Climate Change impacts: South Africa, and the neighbouring regions, would be hard hit. Over the long term, we *needed* a climate solution, and it was in our interests to engage in getting one that worked. The body of our research on impacts was very good, and the unfolding picture, especially when it came to water resources, was accepted as profoundly significant. South Africa had published its first Climate Change communication report to the UNFCCC in 2000[8] (at the time of writing the second report was being prepared and was due in 2010). The report, and many studies that followed, painted a picture of the unfolding understanding of Climate Change impacts for South Africa. The following statement prefaces the communication:

> Potential changes in climate may have significant effects on various sectors of South African society and the economy. The South African Country Studies Programme identified the health sector, maize production, plant and animal biodiversity, water resources, and rangelands as areas of highest vulnerability to Climate Change, and proposed suitable adaptation measures to offset adverse consequences. Two key cross-sectoral adaptation options that link the various sectors are the establishment of improved national disaster co-ordination and management and the raising of awareness on the potential effects of Climate Change.[9]

Clearly South Africa stands to lose greatly if world emissions continue to rise as currently projected. Impacts will be severe, not only for South Africa but also for neighbouring countries, which in turn would have knock-on effects for South Africa. We run a high risk if the really large emitters do not drastically reduce their emissions.

Finally, South Africans know about negotiation, facilitation and consensus building. We are famous for this!

Even though we had a good enabling environment, the idea of a stakeholder-driven knowledge exercise and developing Scenarios of our future around mitigation actions, was a new concept! So it needed some serious motivation. The Department of Environmental Affairs needed a broader mandate, and had to make the case for it.

The 'what if' questions

One way to make a compelling case was simply to ask some 'what if' questions.

---

8 unfccc.int/resource/docs/natc/zafnc01.pdf.

9 It will be useful to refer to the South African Second Communication, which was due to be published in 2010.

1. What if the current logjam at the UNFCCC process finally broke free; after the election pending at the time in the United States broke from the Bush administration approach to a new co-operative one, but demanded deepened involvement from the developing countries with high relative emissions?
2. What if the process stalled and the emissions of others started threatening the South African climate system and we started suffering real damage? Was failure of the process an advantage to South Africa, and in our interests? But equally was success in our interests as well?
3. What if markets moved rapidly to the low carbon economy, leaving us with stranded assets and unwanted high-carbon goods?
4. What if other big developing countries started committing to mitigation actions?
5. What if failing to plan our development and energy pathway in a more coherent way resulted in an energy crisis?
6. What if we were asked to make choices, but we could not due to lack of data and understanding?
7. What if we made choices based on bad data, only to discover later that we could not deliver on our promises?
8. What if we made commitments, but received opposition from local stakeholders?

These questions helped to motivate the case for an 'open ended' study.[10]

First steps

The first step the Department took was to introduce the LTMS idea at South Africa's first large Climate Change Conference, which took place in October 2005. The conference was aimed mainly at an examination of Climate Science and the expected impacts for South Africa, and to expose South Africans to the developing concern on Climate Change impacts. But what was said about mitigation was also notable. Deputy President at the time, Phumzile Mlambo-Ngcuka, pointed out that:

> South Africa must not abdicate responsibility, and that we will do more than we are required to do, mobilising the different sectors in our economy in order to do more.

At the conference the idea of producing a study on our mitigation potential was also put forward to stakeholders. Here is the official statement:

---

10 See Annexure 2.

> At the National Climate Change Conference October 2005: Government, business and civil society concurred that we need to 'initiate a detailed Scenario building process to map out how South Africa can meet its Article 2 commitment to greenhouse gas stabilisation whilst ensuring its focus on poverty alleviation and job creation'.

The support received at the conference paved the way for a full Cabinet mandate to the Department of Environmental Affairs to commission the study. The decision was later translated into this statement by the Deputy President:

> The question is no longer whether South Africa should take on responsibility, but how much responsibility it can take on and by when, and how this action will balance with the development needs of its people. The country will need to ensure that it has the appropriate domestic policy and institutional capacity to implement the outcomes of the next round of negotiations.

At the 2005 South African Climate Change Conference three important decisions were made. The first was an acknowledgement that Climate Change was real and was happening; the science was put on the map. The second was that South Africa would develop a Climate Change response policy through a transparent, participatory and scientifically informed policy development process, providing a focus for action. The third was that a set of mitigation Scenarios would be developed to inform the mitigation components of the policy package, reinforcing the science-policy dialogue initiated at the conference.

The table was set for the detailed planning and implementation of the LTMS project.

Chapter 2

Method

'The future is already here; it is just unevenly distributed.'

William Gibson, science fiction author

'Assessments are not just about the findings. Getting the process right, from the early stages of design through to the communication of findings, is essential in order to have an impact.'

Ash, Bennett, Scholes et al.

There were three primary building blocks to our plan for the Long-Term Mitigation Scenarios (LTMS): Scenarios, Stakeholders and Research (in no order of importance). Each represented a choice, and hence each presented process and content challenges (and headaches!) for both Harald and I. We had decided very early on to implement the government brief using these building blocks. Now we needed a method.

In hindsight the LTMS also has a lot in common with the methodology of Assessments, and I am persuaded through my interview with Bob Scholes of the Council for Scientific and Industrial Research (CSIR), that the LTMS is, in fact, something of a combination of two classic methods. I will accordingly reflect on both, in an effort to explain our approach and reflect on its methodology.

The case for Scenarios

The LTMS was 'about' Scenarios. Scenario thinking was central to its approach. How I got to thinking about Scenarios and their application to the LTMS is a story of a hunch, some careful planning, and some improvisation. I have to say upfront that I am not a Scenario specialist. In fact I have never been totally persuaded by what I see as the results of Scenario processes. But I wanted very much to try the method out in the LTMS, and when we received our brief I was ready and excited to apply the ideas we had developed.

Scenarios are classically understood as 'plausible and internally consistent stories about the future that help organisations and individuals to achieve a broad and open-ended adaptability to inherent unpredictability'.[11] They are usually seen as clusters, often of four such future stories. These are not predictions, but are rather 'what could be'. As a group they are alternative dynamic stories of what could happen. These alternative fictions are described to give institutions a set of future contexts, and with these in mind, they can shape current actions. This is the purpose of Scenario processes.

In the LTMS we essentially re-engineered the Scenario approach but we retained one essential element. This was the potential for creativity that is inherent in Scenarios. The open-endedness, the open agenda, the scope for extended possibilities – call it what you will, this was essential to our approach. Given that in the planning I had based the process on stakeholder involvement, I thought that using Scenarios as a tool for the stakeholder team would give them the space to explore the mitigation challenge. I wanted them to feel free and to feel safe in that exploration.

It was also the intention of the LTMS to provide the South African government with a set of Scenario choices. We had planned a 'menu of possibilities' for the country, and relied on the non-prescriptive approach that this embodies, to bring to the table an open-ended approach.

In short, *we wanted an 'emergent' exploration by multiple stakeholders of various future mitigation pathways, backed up by strong research, which set out some possible options for government.*

Scenarios are about looking far into the future. They are long term. In contrast most organisations plan in windows of 5 to 10 years; governments are no exception. Some investments have long horizons, but often their planning is still based on short-term feasibility. When one looks much further into the future, say over 50 years, more and more uncertainty enters the picture. In the LTMS we wanted to look far into the future, and 2050 was set up as our 'horizon'. We did this because the issue of mitigation is a long-term problem, and the Intergovernmental Panel on Climate Change (IPCC) horizons are typically 2025 and 2050. It seemed prudent to select this date as the endpoint.

> 'You can't do this kind of work without the Scenario approach, without a wide scope, without a learning process, without exploration of all the options. One can narrow down later when one develops the actual policy.'
> *Shaun Voster, former Adviser to the Minister*

[11] Holmgren: *Future Scenarios*, Chelsea Green, p 56.

Scenarios accommodate (at least in theory) high degrees of uncertainty, precisely because they are sets of alternative future stories. One is in effect saying: so far in the future, the world could look like, w, x, y or z – we don't know which one, but we can plausibly describe the alternatives. If each of the Scenarios in the set looks plausible, then they become powerful communicators and informers of present action.

> The Scenario process is logical. There is no place in the core of a Scenario conversation for positions or values. Instead the discussion is about facts and logic: can you convince your fellow team members that the story you are putting forward is plausible?[12]

'Because we were not directly involved in planning, but rather in research, we gained the freedom to look far wider than people would normally have looked. In this way we actually discovered that a lot more was possible than people had thought, that interventions that they said were unrealistic were actually compelling.'
Richard Worthington, formerly Earthlife Africa

Finally, Scenario thinking is recognised as a way to prise participants from firmly held positions. In being asked to imagine uncertain futures, Scenario builders must leave their own context (their own corporate positions, policy or financial constraints) and as individual thinkers consider in an open-ended way how the future might be described.

Adam Kahane describes this aspect in the Mont Fleur Scenarios:[13]

> The third result – the least tangible yet most fundamental – was the change in the language and thought of the team members and those with whom they discussed their work. The Mont Fleur team gave vivid, concise names to important phenomena that were not widely known, and previously could be neither discussed nor addressed. At least one political party reconsidered its approach to the constitutional negotiations in light of the Scenarios.

'LTMS changed things from sectors defending their turf to an integrated response to the problem.'
Kevin Nassiep, CEF

In the LTMS we were keen to see team members freed from their constraints, whatever these may have been. We hoped that the Scenario approach would assist them in doing this.

---

12 Adam Kahane: *Solving tough problems*, www.generonconsulting.com/publications/papers, p 3.

13 Ibid.

But in one particular way we were going to depart radically from classic Scenario practice: we were essentially planning to produce *evidence-based Scenarios*. We were going to use science to build our Scenarios for the LTMS. We wanted these Scenarios to address the high dynamic social complexities that the Climate Change challenge held for South Africa.[14] And we wanted our work to be *systemic* and *emergent.*[15] We would use science, building block by building block, to build the Scenarios (or at least one of them) and then let the future 'emerge'. Modelling was key to this approach.

The idea also developed to use more than one type of Scenario. Some would be based on the modelled approach, some on simple deduction, and some would be story-based. I liked this combination approach, and I think it had a lot to do with the eventual results. It helps, therefore, to explore Scenario process alternatives a bit deeper.

Scenarios in the classic sense, (if there is such a thing) are very useful for creating stories describing unpredictable futures. Communities don't have the capacity to accurately predict the future, and so Scenarios take account of alternative futures in telling what could be. Scenarios are usually presented in groups, as *alternative dynamic options*. They work best as story-based answers to compelling questions about an uncertain future. In fact the compelling question is the starting point. It is the driver for the entire Scenario set.

We did have a compelling question: should South Africa consider mitigation of greenhouse gases in its development planning? Or put more simply, could South Africa afford to develop without concern for mitigation?

Evidence-based Scenarios are very different in nature, because they present options for future outcomes which are based totally on evidence. This evidence is made up of facts and assumptions, which are then extrapolated over time using a tool designed for

'In exceeding its mandate the LTMS became more powerful. Had it just maintained itself in evidence-based Scenarios, it would have been a very scientific but very muted approach to what had to be done. But when it turned into a contextual approach and started presenting the 'what ifs', it started coming up with very real questions regarding South Africa's development options. LTMS allowed the exploration of these development options with more freedom and flexibility. This speaks to politicians' needs; it allows them to come up with their own proposals.'
Kevin Nassiep, CEF

14 Adam Kahane: *Solving tough problems*, p 31 discusses these three complexities.
15 Ibid. p 32.

this purpose, such as a model. Evidence-based Scenarios don't result in sets of alternative stories. They include their own internal uncertainties, rather than becoming sets of alternative stories within uncertain future contexts.

I consulted with three Scenario experts: Clem Sunter, Louis van der Merwe and Adam Kahane. All three had had extensive experience in this field. Clem Sunter had done recent Scenario work on Climate Change in London, and presented this work to us, using his 'quadrant' approach (more about this later); Louis presented some compelling thoughts about the 'take-up' of developed Scenarios by the broader audience; and Adam had done pioneering Scenario work internationally, and was personally very interested in the Climate Change problem.

What Adam, Clem and Louis helped me to understand was that we needed to adapt the 'classic' Scenario approach for the LTMS project. There seemed to be a need to combine the type of work they had done with our intended approach to evidence-based Scenarios, which would be based on (scientifically) modelled futures. What we needed, in fact, were Scenarios 'within' Scenarios.

Our plan was to develop both the evidence-based emissions trajectory Scenarios, and some alternative external context-driven Scenarios against which the evidence driven ones could be evaluated. This was really central to the LTMS method, and it was the 'heart of the approach'. Had we only presented the evidence-based Scenarios as a set of options, there would have been no value-driven evaluation of them. This is what the story-based Scenarios eventually provided, and through them we achieved a remarkable result.

On their own the evidence-based Scenarios were very compelling too: they contained and were built on a detailed and voluminous set of inputs, all to be agreed by the stakeholders. This component-by-component approach (in which you work on all the smaller pieces of the puzzle first and then pack them together afterwards to see a result) turned out to be very persuasive.

We planned to use a Scenario 'game' at the beginning of the process to introduce our new stakeholder team to the Scenario approach, and to our compelling question about development and mitigation. This would be our kick-off point. We would then go on to construct our evidence-based Scenarios. In the end we added a final Scenario exercise to evaluate our evidence-based Scenarios, and to test the robustness of our science-based results against possible external contexts, both national and international. So ultimately there were essentially three distinct Scenario processes in the

LTMS: Scenarios examining the context in South Africa today (with reference to growth and emissions); evidence-based modelled Scenarios (the main body of the LTMS); and Scenarios of the world in 2050 (the stories that tested the robustness of the modelled Scenarios).

The heart of the Scenarios in the LTMS was to be the evidence-based mitigation trajectories. Some reflection on how these were conceived is important.

Mitigation is a complex and dynamic process, occurring on multiple fronts and taking place unevenly over time. The reduction of greenhouse gases is either conscious (such as actively introducing measures), or it occurs as an unintended benefit from another activity. It can be borne in policy, or through investment; it can be purely behavioural in nature, or emerge over time through research and development of new technologies. But one thing is sure: mitigation takes place in a complex matrix and planning a mitigation path is certainly not simple.

In the LTMS we did not have a single set trajectory (we were setting alternative ones), and accordingly our starting point could not be 'here is the trajectory, and there are four possible ways of aligning with it'. In the LTMS we had to take one step before this, and actually set up a few alternative trajectories, situated within an 'empty policy space'. So we set out to produce a few possible trajectory options, and each one would be a collection of mitigation actions. Our premise was: 'what possible overall routes may notionally be available to South Africa, and what is each one's value and impact.'

Scenario fundamentals

Scenario building can be carried out following different approaches. The central idea in Scenarios is:

> The assumption underpinning the methodology of Scenario planning is that key elements or determinants of the future are present in the current reality, and thus there is a scientific method – a futures methodology – to access and expose them.[16]

There are a few possible alternative 'futures methodologies' for unlocking Scenarios. Among these are:

---

[16] Van der Heijden, 1996, http://www.informaworld.com/smpp/title~db=all~content=g916779887.

- **Normative Scenarios:** these start with a preliminary view of a possible future, and look *backwards* to see if and how this future might or might not develop out of the present reality. This approach would clearly not work for a Scenario resting on data; but in fact our second Scenario in the LTMS was a Normative Scenario to some extent. This will be seen later in more detail, but even that Scenario ('required by science') was built on a future fact set, and roughly extrapolated backwards.
- **Exploratory Scenarios:** these start with the present as a starting point and move forward to the future by asking 'what if' questions about implications of possible events outside familiar trends. To some extent we used this approach at the beginning and at the end of the LTMS to examine our Scenarios against external contexts, themselves Scenarios. This will be explained in some detail. These were the Scenarios within which our evidence-based Scenarios were situated.
- **Inductive (or bottom-up) Scenarios**: this approach builds step-by-step on the data available and allows the structure of the Scenarios to emerge by itself. The overall framework is not imposed, and the storylines grow out of the step-by-step addition of the data. We used a hybrid of this approach, especially in the 'Growth without Constraints' Scenario. We built the Scenario by taking as much information as possible (in our case, modelled data) into account. The Scenario emerges on its own without being imposed at the start in this method. It produces a trajectory rather than the description of a future 'point'. This was quite crucial in our process, as will be seen later.

> 'My learnings from being deeply involved in various Scenario processes are that there isn't one correct way of conducting such thought processes. There are useful archetypes and previous Scenarios that can simply be taken and tweaked as well. You don't have to start completely from scratch.'
> *Bob Scholes, CSIR*

- **Deductive (or top-down) Scenarios**: pieces of data are fitted into a framework. We used this approach for what we called the 'Options'. These were developed as alternatives to our main Inductive Scenario. This was a relatively simple approach.

In summary, our approach was to run four stories (Context Scenarios) as an opening scene-setting process, followed by a set of two main Scenarios (both based on science) with some Options within them, followed finally with some evaluative stories.

I shall deal first with the Context Scenarios.

'Context' Scenarios in the LTMS

The two sets of 'Context Scenarios' rely on a specific set of theoretical foundations:

> We can extrapolate ... drivers of the future and apply both qualitative and quantitative methods to investigate them, in order to intelligently discuss what may someday happen in a more informed manner. However, different people hold different opinions about the key drivers of different future scenarios, and hence the methodology has to accommodate as many differing assumptions about the present as possible. This is an important aspect, as the implication is that future Scenarios involve exposing and comparing different sets of assumptions about the present, which then can be inferred to the future.[17] In this way Scenario planning or futures methodologies are as much about the 'here and now' as they are about any distant future; in fact, future Scenarios may be viewed as an intelligent but playful way to use the future as a vantage point from which to view the present. Accordingly, this becomes simply a radical way to discuss the current reality.
>
> Importantly, Scenarios are not the same as forecasts. Forecasts are based on a single understanding of the present, which is then extrapolated to the future. Scenarios on the other hand are based on *different* assumptions about the present, which are then extrapolated to different futures.[18]

In order to capture these different assumptions, a useful method is to use the 'quadrant' approach. In his approach, the Scenario facilitator will have conducted research and core interviews to procure a pair of *key variables that may possibly influence the future*. These variables are plotted on 2 × 2 axes, and then different Scenarios are extrapolated from the resultant matrix. Quantitative data is then fitted into the framework.

Clem Sunter's diagram at the top of the next page represents the basic approach.[19]

---

17 Van der Heijden, Bradfield, Burt, Cairns & Wright, 2002.

18 Pieter Fourie, Department of Politics, University of Johannesburg, *African Journal of AIDS Research* 2007, 6(2): 97–107.

19 Clem Sunter. 2001. *The Mind of a Fox.*

The Clem Sunter/Chantell Ilbury matrix from *The Mind of a Fox*

High control

3 Options

Given the scenarios, identify what the implications are and what options are available.
Complex integrated model simulation optimises demand and supply options and generates results for consideration. Sensitivity analysis explores the depth and implications of potential options.

4 Decisions

Given a deeper understanding of the relevant knowledge and potential for change there is now a wider selection of options available. This helps to make better decisions and develop more robust strategies and policies.
The proposed energy plan can be compiled. This should be accompanied by a macro-economic impact assessment.

Uncertain — Certainty

Identification of uncertainties, driving forces and trends. Prioritisation of key drivers to be used to create scenarios around "alternative futures".
For example, impact of climate change, international initiatives to mitigate against climate change, resource estimates (cost and volume), price volatility, security of energy supply and imports, international and RSA economic growth, trade agreements, global cooperation, technology developments etc.

2 a) Key uncertainties
b) Scenarios

These are the things that we know about with a reasonably high degree of certainty. We don't have much control over them though. They are the pre-existing facts, knowledge, rules, laws, processes, stakeholders, etc. Definition of the system being focused on.
For example, existing energy balances, resource estimates, technology options, policies and legislation. Past trends on energy demand, industrial development, markets, GDP and population growth.

1 Rules of the game

Low control

The example at the top of page 26[20] illustrates the results of this approach: four key variables are agreed, and transferred to two axes in this Climate Change Scenarios case.[21]

The effectiveness of Scenarios

If we had only used these 'contextual Scenarios', we would only have been able to present South Africa with a few broad approaches, helpful, but not nearly empirical enough. Everyone would have liked the 'clean, green, competitive' Scenario, for example, but no-one would know how to *get*

20 www.mindofafox.com/casestudies.php.

21 For more illustrations, see Annexure 3; for a note on the basic method of producing these quadrant Scenarios, see Annexure 4.

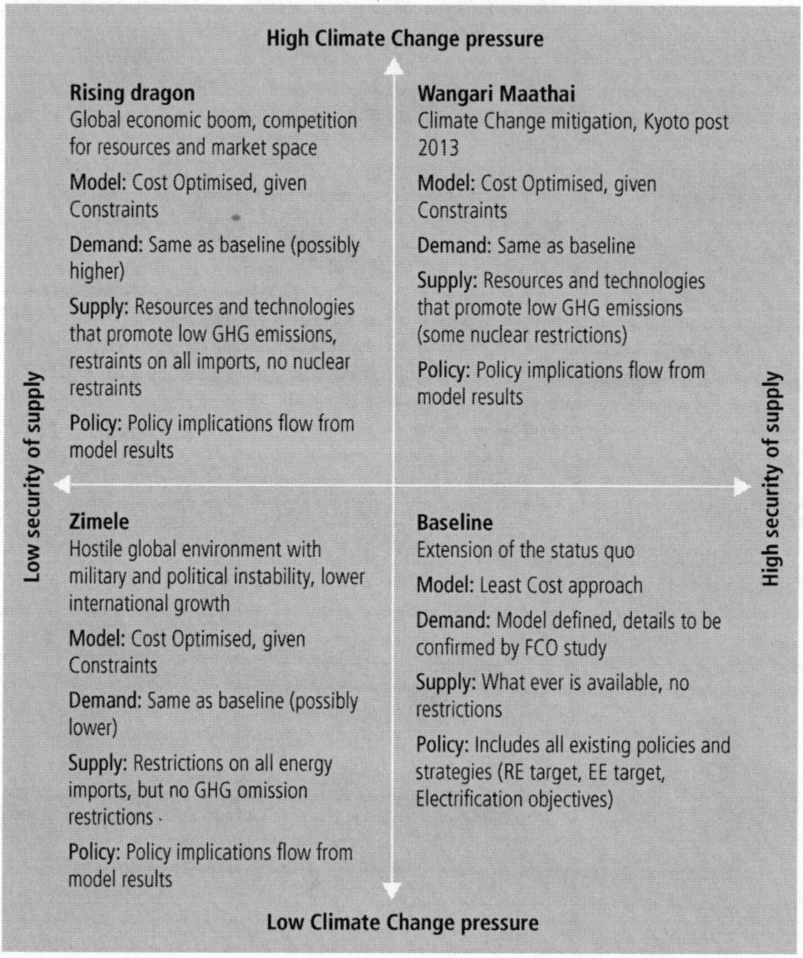

High Climate Change pressure

Low security of supply ← → **High security of supply**

Rising dragon
Global economic boom, competition for resources and market space

Model: Cost Optimised, given Constraints

Demand: Same as baseline (possibly higher)

Supply: Resources and technologies that promote low GHG emissions, restraints on all imports, no nuclear restraints

Policy: Policy implications flow from model results

Wangari Maathai
Climate Change mitigation, Kyoto post 2013

Model: Cost Optimised, given Constraints

Demand: Same as baseline

Supply: Resources and technologies that promote low GHG emissions (some nuclear restrictions)

Policy: Policy implications flow from model results

Zimele
Hostile global environment with military and political instability, lower international growth

Model: Cost Optimised, given Constraints

Demand: Same as baseline (possibly lower)

Supply: Restrictions on all energy imports, but no GHG omission restrictions

Policy: Policy implications flow from model results

Baseline
Extension of the status quo

Model: Least Cost approach

Demand: Model defined, details to be confirmed by FCO study

Supply: What ever is available, no restrictions

Policy: Includes all existing policies and strategies (RE target, EE target, Electrification objectives)

Low Climate Change pressure

there, or whether there was more than one way of getting there, and how much it would cost, and so on. To be effective, such Scenarios must be *plausible*, *consistent* and offer *insights* into the future. Readers must be able to say: 'yes, I believe that could happen, it makes sense, and it relates sensibly to current realities'. But in our case it was the roadmap that concerned us, not just the destination.

We also wanted the government to make *use* of the Scenarios, not just 'note' them. Hence we knew that they would have to have a decision-making utility. Our Scenarios had to do more than simply convince people of these possible future stories – they had to be a tool to help actors, and our government in particular, to make decisions. Scenarios only really have a purpose if they can help leaders to think in a disciplined way about the future when making decisions. The method ideally helps the decision-maker to consider the range of plausible futures, to articulate preferred visions of the future, and to use what is learned during the Scenario development process in the formal decision-making process. It also helps to stimulate creativity in the leaders who participate in the process and to break from the conventional obsession with present and short-term problems.

Ultimately we opted for a combination of Scenarios as I have described. In the first Scenario meeting, we would explore a general contextual set of Scenarios, and place Climate Change and Development centre stage; in the following meetings we would build the modelled Scenarios. We were clear from the start that we wanted to combine the freedom of contextual Scenario thinking with the compelling nature of science-based trajectory-type Scenarios.

The assessment approach of the LTMS

> An assessment will have the greatest impact where consideration is given to both process and products, where stakeholders are fully engaged, and where assessment design follows scoping of user needs.
>
> *Ash, Bennett* et al.[22]

The science-based Scenarios were typical of assessment processes. During the LTMS process, we never used the word 'assessment' once. However in many ways the LTMS was in part an assessment, and the principles apply to both types of processes. Bob Scholes and his colleagues, Neville Ash and others, describe the process of developing assessments.[23]

In the exploratory phase of an assessment, the *need* for the process is determined:

> The concept of an authorising environment is a useful way to ensure that an assessment has the necessary level of buy in from key stakeholders. The authorising environment is the set of institutions and individuals who see an assessment as being undertaken on their behalf and with their endorsement and engagement. Examples might be village elders, land managers,

[22] As yet unpublished: all quotes below from Chapter 1 of the draft: full recognition is owed to the authors.

[23] See Annexure 5 for some examples of assessments cited by the authors.

> agricultural cooperatives, or local or national governments. In practice, whether or not the members of the authorising environment have provided formal authorisation, the true test of whether the authorising environment was sufficient is whether those stakeholders have a substantial ownership in the final products and a commitment to take actions based on the findings.

In many ways this paragraph says it all: in the preparation of the LTMS we had our compelling question, and our authorisation; we now needed the buy-in, the 'substantial ownership' and then we could hope for eventual 'commitment to take action based on the findings'.

The methodology for assessments described by the authors is also entirely appropriate for the sort of exercise the LTMS was intended to be.[24]

Achieving social accountability through stakeholders

All planning, and all imagining of the future, is ultimately done by, and about, people. Hence our second primary LTMS building block was the stakeholder element.

> 'The LTMS showed us the difference between what was desirable and what was practically possible.'
> *Kevin Nassiep, CEF*

Large groups of opinionated people tend to like endless argument, and it's well known that you can't 'draft in committee'. It is much easier to generate plans without the involvement of groups. The problem is that the people who are affected by them, especially those who are required to carry them out, generally reject such plans.

For the LTMS, we wanted a result that was supported by stakeholders. We wanted this for a number of somewhat obvious reasons. In the first instance industry stakeholders would have to make the primary mitigation effort, and hence they would have to believe in the study, support it and carry out whatever policy was to be based on it. Secondly other sectors would have to support the plan, as it would have impacts on jobs, poverty alleviation and other matters of interest to civil society actors. Finally, for government, we wanted a sense that policy outcomes had support; conversely we wanted government to be encouraged to act in a way that was socially accountable to a plan that had broad-based civil support.

However, there is a big 'but'. Working with stakeholders is risky, and it takes time. We had no idea if the stakeholders would agree to the Scenario

24 See Annexure 6 for some useful excerpts.

process, or to the detailed approach, or to any of the inputs or assumptions. We had no idea if they would agree to anything! We certainly knew that there would be huge differences of opinion. How would we hold it all together? How would we keep the dinner-party peaceful – and focused?

I felt right from the start that the proposed LTMS process would be characterised by:

- High levels of complexity, especially in terms of technical information;
- Many stakeholders, depending on the size of our Scenario Building Team;
- The presence of conflict, over such issues as nuclear versus renewables, and maybe over the Climate Science;
- Capacity differences, between those familiar with the technical issues, and those who are not;
- Strong vested interests, such as in the dominant coal industry;
- Value differences, between civil society leaders and industry.

This is where facilitation can play a profound role, and indeed the LTMS was planned around third-party facilitation. We effectively turned to third-party involvement as a way to hedge the risk of conflict. This meant a large responsibility on the facilitators.

Facilitators find the routes towards the mental and emotional places where people want to make deals. Facilitators are 'activist' in their approach, and are the careful designers of process pathways within which stakeholders operate. Facilitators use many tools to achieve this: through caucus mediation, by creating doubt, through working groups, bilaterals, and so on. In short, facilitators have to become ultra-creative, which they must do behind the scenes in both the preparation and design of processes and in the interventions themselves.

> 'We explored the unexplored.'
> *Imraan Patel, DST*

Facilitation in the LTMS

I worked with two other facilitators at various points of the LTMS, but my role was really to hold the entire process together. I do not want to embark on an extensive treatise of the facilitation components we applied in the LTMS, but here are some of the cardinal ones:

1. In the first instance we planned the LTMS as a process with three phases: the Scenario Building phase, an evaluation phase by high level leaders, and finally a political phase (where the government would react to the LTMS study and determine the next steps). This layering of the process was very important for its success.

2. Across the processes and within them, facilitation is a way of ensuring *management* and *momentum*. As process manager, the facilitator works closely with the technical team (which essentially produces the raw material making up the evidence for the Scenarios). This 'driving of the process' includes everything from planning the process milestones to the approaches that will be taken inside the meetings; from making sure parties are heard and at their ease, to managing the data and record-keeping. So this is in a way the first and perhaps most important skill: holding the ship on course.
3. I also took on the task of ensuring *transparency*. Information exchange is a large part of these processes and information can become voluminous and dynamic in nature. If this is inaccessible to stakeholders, or is perceived to be accessible to some stakeholders but not others, the process will fail. The facilitator is a *knowledge manager*. I did a lot of work on presentation material, and issued a good few process notes, so that people could see where the roadmap of the process was leading us.
4. Facilitation is a slow and emergent process, and is all about *alignment* of stakeholders' interests. The facilitator is watching this element above all else: where and what are these interests, how do they impact on the process, and how are they changing? Strong stakeholders may want quick fixes that suit them, but the facilitator must slow down this imperative, and focus on the slowly unfolding decision of the group, not powerful individuals. Facilitators must watch the timing. The rule seems to be: too fast and stakeholders will suddenly feel that they have been manipulated and will want to go back; too slow and everyone will lose momentum and interest. The pace of the LTMS was not rushed, and this was a good thing.
5. Facilitators must focus on the *result*: after all eternal consultation is not the point. The road map of the process needs to be revisited constantly, and it helps to know when one has reached the markers. Harald and I pushed for closure on issues all the time.
6. I believe that processes that look to the future, such as change processes, Scenario processes and the like, need some *visioning* from the facilitator: he or she should be allowed to help stakeholders look out of the present towards what possible futures may look and feel like. The facilitator needs a special licence.
7. Facilitators must keep it *light*: long multi-party processes can be gruelling, so some laughter and light relief counts. Pascal Moloi, who facilitated the High Level segments of the LTMS, was brilliant at making

people feel 'in the process' – it helped that he knew many of them and was charming as well.

8. Facilitators must *know the topic*: some knowledge gaps are not a problem, but a facilitator who is essentially lost when it comes to content cannot be a help to the stakeholders. This is true especially in processes that are content-heavy, such as the LTMS.
9. Finally, and very importantly, is the business of process *design*. In the LTMS we spent as much time planning the process (mostly Harald, Pierre and myself, with the management team checking our plans and playing a strong reviewing role). This preparation is as important as it is obvious. In fact, I would say that 90% of the facilitation of the LTMS occurred in behind-the-scenes preparation and process design work.

It remains obvious to add that facilitators must be independent and non-aligned to interests within the stakeholder group. In my case I have worked extensively with all sectors in South Africa on Climate Change, and so most were familiar with me and knew that I held no bias against any sector or institution. I also kept my views on some of the big issues to myself! I am outspoken on the fact that we need to take bold steps to resolve the Climate Change crisis, but I am also open-minded on solutions, and very aware of how tough the challenge will be. In that spirit I am happy to work with coal-mining interests one day and climate activists the next.

'To my mind the LTMS was a very participatory process. Participation does not mean that every person needs to sign off on the study. The South African notion of significant consensus is here a very useful one.'
Bob Scholes, CSIR

Consensus seeking

We were committed, right from the start, to consensus-driven inputs to the modelling. In the LTMS we approached this problem as follows: we agreed that consensus (which included the concept of 'sufficient consensus') was imperative for all inputs and assumptions, and where agreement was not

'Why was the LTMS a success? Firstly it was designed as a process that suited the South African culture of negotiation. Then it was multi-layered: Deep analytical and technical work in the SBT, with that information then translated by sherpas to the leadership, then the Minister and the leaders, building consensus at a high level. There were three levels. Then you also had the four stakeholder groups in the same room. Then you also gave it 2 years to mature: had we rushed it, it would have failed. These factors added to its success. As a nation we have a culture of inclusivity.'
Shaun Vorster, former Adviser to the Minister

possible, to omit split positions and if we could, move on without such inputs. This meant that some data was compromised, and some just left out. We agreed that this was better than a result that was not supported by some stakeholders. We also agreed to deal with all technologies even though we knew that some stakeholders did not in principle support them.

A word on social accountability

Most of the critique of the LTMS falls under the banner of social accountability: accusations that the process was not accountable enough; that participation was skewed in favour of industry; or that a wide reckoning of the views of South Africans in general was not taken.

These criticisms would have been true had the LTMS been intended as a decision-making process. However it was not. Policy-making that is accountable to the people demands a much wider input from civil society than was the case in the LTMS. The LTMS was a study, which was conducted by a group of experts who also happened to be stakeholders within the various sectors of society. As a study, should it have been more accountable to the sectors in society? Should the results of the study have been subject to deeper, more representative interrogation by society?

I personally think that studies stand or fall by their results, and that they don't need (in most cases) to be subject to wide survey before being presented as legitimate work. One could say that the group of experts who produced the study were not qualified for the task, or that the output is inferior; but there is no requirement for full representative consultation before a study is seen as legitimate.

To my mind one form of social accountability takes place when government makes decisions or policy whilst taking into account input from civil society (in the broad sense of the word). The South African government reacted to the LTMS study quite quickly after publication, and this reaction did have its supporters or retractors.

The study on the other hand must be gauged on its content, not on who was included and who was not.

> 'The LTMS is policy informative and not policy prescriptive.
>
> Deeper participation plays a role during the policy making process and not during the assessment necessarily.'
>
> *Bob Scholes, CSIR*

It must be added that in designing the process, we were conscious that the LTMS in its initial stages would present a great deal of complexity, and that participation at this stage had to be limited. Again this shows that complex issues need to be shaped in stages of accountability. In the

LTMS we went 'wide' (the 2005 Conference), 'narrow' (the SBT), 'wider' (the High Level segment) and then 'wide' again (the 2009 Conference), which illustrates one way of balancing the imperatives of accountability and effectiveness. Where I think we could have improved on the range of social involvement was in the High Level phase, where the audiences proved rather select and small. There could perhaps have been deeper input, for example, during the 2009 Conference, but by that time Cabinet had already responded to the study. And of course, one could have taken the LTMS concepts to the public for a deeper engagement and understanding.

> 'This was a very technical process, and some members of civil society organisations were less than equipped to participate meaningfully. It was not an easy process. There was a lot of documentation. This type of process presents specific challenges for civil society. But in the LTMS civil society engaged in the robust debate in the SBT, and was very vocal.'
> *Richard Worthington, formerly Earthlife Africa*

The LTMS as a study is a very technical exercise. The results do not make for bedtime reading. Nonetheless, the 'LTMS' has become something of a household name in South Africa and as such it has worked well in generating a degree of social awareness of the issue of Climate Change. It has had wide media coverage and been discussed in many boardrooms. It shifted the debate from the science to the issue of building a new green economy. The Budget Speech that followed LTMS included, for the first time, multiple references to the low carbon economy challenge for South Africa. In 2010 Ministers and the President made the case for the Green Economy. I think it is fair to say that these developments all had their origins in the government's decision to conduct the LTMS study.

Finally

These foundations of the LTMS process (Scenarios, the assessment approach, facilitation and involvement of stakeholders) were then coupled with best-practice research. This combination was key to the plan that we then unfolded and applied.

> 'How do you build policy instruments in the same way that LTMS built a story? How can we build evidence-based policy? This is an interesting challenge.'
> *Imraan Patel, DST*

In the next chapter I will reflect on the Scenario-building process itself. Our task had been set, and our outcome described:

South African stakeholders understand and are focused on a range of ambitious but realistic Scenarios of future climate action both for themselves and for the country, based on best available information, notably long-term emissions Scenarios and their cost implications.

Chapter 3

Building the Scenarios

'Scenario development is "the gentle art of reperceiving".'

Pierre Wack, who pioneered Scenario planning at Royal Dutch/Shell

Chronology of the LTMS study at Scenario Building Team (SBT) level

SBT 1 16 August 2006

I am always nervous of the beginning of process interventions. Starting processes is often the most difficult part. People are understandably suspicious of new processes, and cautious to lend their time and stake their reputations on something that may fail miserably. The opening phases or steps are the most difficult and I have seen many processes heading for early shipwreck due to bad decisions in the opening phases.

Assessment processes, as Bob Scholes has helped me to understand, have particular social dynamics and to be significant and 'successful' they require three key elements: *legitimacy* (in the authorising environment); *relevance* (addressing a question that interests society); and *credibility* (engaging scientific stakeholders in a way that is defensible).[25] These three requirements are also the preconditions for successful processes, and must be achieved before implementation starts.

In the LTMS we had legitimacy, due to the process having been authorised by government. I cannot emphasise enough the importance of this, especially the fact that the instruction had come from the entire Cabinet and not only one department. This powerful mandate would ensure that stakeholders came to the table, at least. They would want to be there. It showed when I called stakeholders to the Scenario Building Team (SBT). The Cabinet mandate put matters beyond question.

---

[25] See Bob Scholes' work in this regard.

We also had relevance, in the sense that the question being posed was compelling and important to both government and non-government actors.

Now we needed to ensure credibility. Credibility for the LTMS would mean selection of the best stakeholders for the process, as well as the appointment of our best researchers.

Setting up the SBT

The LTMS would have two distinct phases, the first being the technical work, and the second the presentation of the work to South African leadership for a wider mandate. In the first phase we would form an SBT made up of stakeholders, to drive the technical work. These stakeholders would work in partnership with the Research Teams and develop the material.

My initial task was to determine a methodology for the selection of these SBT members, and to find them and recruit their services (for free!) to the process.

We presented a strategy for doing this to the Management Team.[26] I suggested that we could solve the difficult problem of SBT appointments by avoiding selection by organisation or sector (representative selection), and by focusing instead on individual expertise within broad sectors. We opted for a medium-sized group of individuals, drawn broadly from all the relevant sectors and organisations, and chosen for their high level of technical skill. We had two criteria: skill and involvement in one of the main sectors.

The identification of these sectors in industry was relatively simple (there is guidance from the Intergovernmental Panel on Climate Change [IPCC]), but we also needed to take an organisational approach, given the large emission contributions of two organisations, namely Eskom (South Africa's only utility) and Sasol (the large coal and gas to liquids operator), as well as dominant energy users, represented by, for example, the Energy Intensive Users Group (EIUG) and the Chemical and Allied Industrial Association (CAIA). We also had to look at where the emissions were concentrated within the sectors themselves – sectors such as cement production, paper and pulp, mining and so forth. Our selection included a blend of all these interests and particular characteristics of the South African sectoral landscape. In other countries this would no doubt have been a very different picture.

---

26 See Annexure 7 on the structure and composition of the LTMS management team.

In the case of government, the Department of Environmental Affairs helped to identify experts from all of the relevant government departments: all in all about a dozen government officials would join their industry colleagues in the SBT.

South Africa has a well-organised labour set up, with most organised labour belonging to one of two federations, namely the Congress of South African Trade Unions (Cosatu) or the National Council of Trade Unions (Nactu). We approached them to put forward experts.

From a selection point of view, our biggest challenge was civil society, purely because of its size and diversity. Civil society was crucial to the process in that its role would be to challenge many of the assumptions put forward by industry. Civil society organisations involved in Climate Change in South Africa range from those for whom it is a specialised function, to those who cover a range of community and development issues as well. How to choose? We needed those who had the ability and the skills to play a technical watchdog role, but there were very few such technicians to be found within civil society. Fortunately most organisations that focus on Climate Change are associated with the South African Climate Action Network (SACAN), within which there were a few people who would be able to participate in the technical work that would follow. We also included some key 'non-aligned' academic and research-based climate leaders, who proved to be invaluable.

The management team suggested that locating and selecting stakeholders for the SBT would be through a one-on-one consultative process, which I would lead. We went through a list of sectors and organisations, including some of the large emitters, and I was tasked to make the approaches. I conducted about 40 approach meetings over a month or two, and in this way built up the list of names for the SBT. I got to meet many new faces as I asked questions such as 'who knows the most about carbon emissions in the cement industry?' I was amazed at the generosity of people willing to contribute to this new study.

Size was the problem. I could easily have amassed a group of 100, but we really wanted only about 40 or so SBT members – at least originally. Ralston and Wilson point out that a Scenario team, in classic Scenario processes, has three primary responsibilities:

1. To define the critical uncertainties (the input assumptions and data);
2. To develop the future Scenarios that cover the key alternative outcomes to these uncertainties (packaging the options);

3. To initiate the process of thinking through the strategy implications of these Scenarios.[27]

The authors point out that the most effective team comprises eight to twelve people. In classic Scenario plays, the team is often quite small. The Mont Fleur Scenarios had a team of 22. This is what Adam Kahane says:[28]

> The team needs to be:
> - *Respected* – composed of leaders who are influential in their own communities or constituencies. They need not hold 'official' positions.
> - *Open-minded* (in particular, not fundamentalist) and able to listen to and work with others.
> - *Representative* of all the important perspectives on the issues at hand. Any stakeholder must be able to see their point of view represented by someone on the team, though they need not be formal representatives of these groups or positions.

It was clear to me that having the luxury of a small team was out of the question. There were simply far too many relevant players. This was especially so in the civil society group. We eventually decided that between 60 and 80 people should constitute the SBT. Government also wanted a number of observers who could build their capacity by watching the process. The room was going to be full.

When it came to final selection, the management team agreed on the member selection for the SBT. Our focus was on people who in the view of the steering committee were the 'national experts'. We believed that we had a credible group.

> 'You can't conduct a highly complex technical process with a stakeholder body that is too large.'
> *Laurraine Lotter, CAIA*

In summary, when making up the SBT, we were driven by the following principles:

- the SBT was to be made up of selected individual leaders;
- they were to be chosen to present strategic thinking, informed by sector views;
- they did not need a mandate to represent their sector;
- their inputs were to be in their personal capacity;

27 Ralston and Wilson: *The Scenario planning handbook*, Thomson, p 69.

28 www.generonconsulting.com/publications/papers/pdfs/Mont%20Fleur.pdf, p 7.

- their inputs would be protected;
- no proxies or replacements etc. were to be allowed unless by arrangement.

I include the names of the SBT members in recognition of their contribution to the LTMS.[29]

The issue of facilitators and process administrators

Other countries and players may find it interesting that professional third party facilitators ran the LTMS process. This was a conscious decision, but did it add value?

We agreed at the outset to ensure that the 'facilitation team' should be made up of facilitators as well as a process administration service. The latter would minute the meetings and administer matters such as venues and communications. As I was an accredited facilitator of an organisation known as Tokiso (www.tokiso.com), they would provide this latter service. As it turned out, the staff of Tokiso struggled at times to record the highly technical debates that would take place, but remarkably given the technical brief, they kept the process administration together and proved to be an important component of the large and difficult to manage process as it unfolded.

I was the lead facilitator of the SBT process. I was initially joined by Edwin Mohlalehi (who later took up a new post as a judge) in some of the SBTs. Later in the High Level meetings Pascal Moloi took on the lead facilitation role.

Apart from the facilitation of the meetings of the SBT and its working groups (and the mediation between members in areas of conflict) I was also in charge of 'process design', which is really about making choices about how the process should unfold. I shared this role with Harald and in this way we made sure that the process and the research were properly combined.

The facilitation of the LTMS was exhausting and, dare I say, the wild horses were not always under control. Harald also ended up having his fair share of surprises and frustrations in the research component. In the end I relied heavily on him for support (and vice versa), and I took a lot of 'time-outs' during the process to check my approach with him. In hindsight this learning approach may have added to the creativity within the SBT process. I was by no means an experienced Scenario facilitator, but I did have an open mind and wanted to be as creative as possible.

---

[29] See Annexure 8.

What is important is that we were a team: facilitation, process design, technical work, presentation and process administration were all incorporated in the facilitation team, which I coordinated in an informal way. With good government oversight we were mostly on track. Harald and I met weekly or more often and we spent as much time on preparing our next steps as we did actually running the meetings. Throughout the process we helped each other, and checked each other's decisions and progress. We also included other ERC staff in this planning process and got invaluable input from Pierre Mukheibir and Andrew Marquard. We struggled at times, and there was a fair share of despondency and harsh words. This is the way of such processes – they are by no means easy.

In short I strongly believe that processes such as the LTMS should be facilitated, if only for one cardinal reason: they have a potential to derail and descend into conflict and terminal disagreement. And facilitators (at least good ones) know how to get them back on track. However, the challenge is to find facilitators who have enough knowledge of the territory to add real value or to find experts in the field that have the makings of good facilitators. Mitigation is a challenging field and most generalist facilitators would be clueless. Most climate experts, on the other hand, are too aligned to specific sectors. I was lucky: I had had almost six years in the Climate Change world, and ten years as a generalist third party practitioner; that said, I was to learn as never before.

Technical content

Four research teams would produce the data. These were the Climate Impacts Team, the Energy Emissions Team, the Non-Energy Emissions Team, and finally the Economy-Wide Analysis Team. (In other countries processes would possibly have different teams: in South Africa most emissions come from the Energy Sector). Our teams had to be made up of the best researchers South Africa had to offer. The work had to be respected and credible. It is interesting that we chose to do all the research with South Africans; this was truly a project for South Africans, by South Africans.

It was essential that the SBT have total confidence in the ability of the research teams to produce unassailable and independent data. The researchers needed to be independent and competent. For the sake of transparency we invited some of the cardinal players (who themselves were doing emissions research, notably Eskom) to insert their own technical experts into the teams.

Researchers at this level were new people in my process life. We all met at a guesthouse in Pretoria before the process started in a 'get to know you session'. On an impulse I thought it would be a good idea for all of us to tell our life stories, which we did to much amusement and amazement. I liked the eccentricity of some of the researchers and the uniqueness of their various stories. In hindsight I am pleased that we did this, because we forged personal connections, which stood me in good stead in dealing with the teams as the process unfolded. They probably thought that choosing to be a process manager was the oddest form of masochism!

Communication

We intended from the start to team up the research groups with the SBT in order to 'build' the Scenarios. Accordingly the next task of the management team was to develop a coherent communications strategy to ensure that all the players, both in the SBT and in the research groups, knew what was intended. A brochure and presentation were prepared, and a special website was created where all work in progress would be accessible to participants. We also set up video linked management meetings.

Wrapping up the preparations

By March 2006 our preparations were complete. Our structure was set and can be represented as follows:

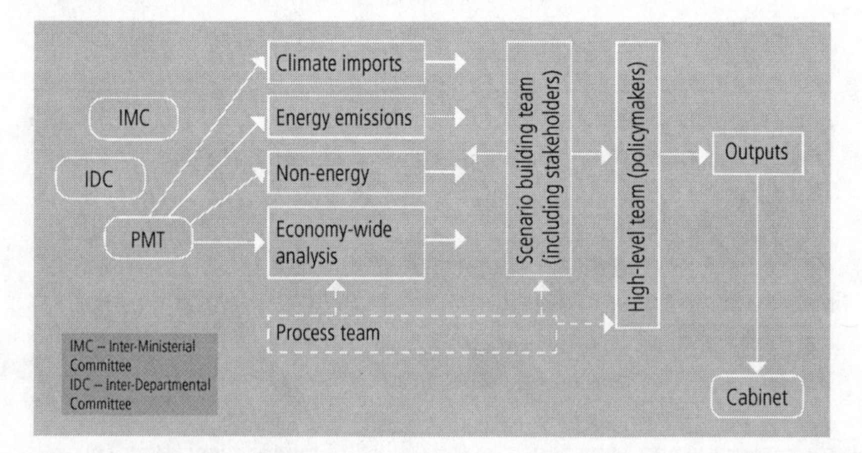

SBT 1 kicks off

The SBT saw us setting up a large u-shaped table in a hotel conference room in Sandton, Johannesburg. I watched as all the people I had interviewed filed in. There were warm greetings, and glances to see who else was there. It was perhaps not the first time all the thinkers in the field found themselves together in one room, but it was the first time they would work together towards one objective. I was quite apprehensive as Harald and I started the process. It was 16 August 2006, a full year of planning and thinking had elapsed since receiving our initial mandate, and a fine spring day.

We had decided to drive the first meeting 'from the front', using a PowerPoint presentation as a framing agenda. Our agenda items for the meeting were:

1. Introduction
2. First session: Process and information issues
    - Roles and players
    - Research component
    - The SBT
    - Status of output
    - Rules of conduct

    Second session: Process and information issues
    - Agreeing on terms
    - Setting the Scenarios
3. Context Scenarios: Group exercise

After opening the meeting and introducing myself, Joanne Yawitch of the Department of the Environmental Affairs and Tourism (DEAT) gave an opening address which set the objectives of the LTMS:

- To provide South African government and other policy makers with a detailed report on Scenarios and mitigation for future Climate Change action;
- To build overall capacity, through this forum, by understanding the issues related to Climate Change and the relationship between our economic growth, our economic growth projections, emission Scenarios and our overall growth development;
- To ensure South African stakeholders understand and are focused on a range of ambitious but realistic Scenarios of future climate action both for themselves and for the country, based on best available information, notably long-term emissions Scenarios and their cost implications;
- Based on this fact the Scenario Building Team (SBT) will engage in a mitigation Scenario planning process to produce a number of possible Scenarios around Climate Change;
- The material will be taken to Cabinet for debate and suggestions, so Cabinet can approve a long-term climate policy and for the negotiation team to have detailed position internationally;

> 'The strong political leadership was cardinal to its success, as was the bottom-up participation and analysis.'
> *Shaun Vorster, former Adviser to the Minister*

- Cabinet policy based on the Scenarios will assist future work to build public awareness and support for government initiatives;
- The LTMS will report to the inter-ministerial committee twice annually, and will have links to the National Climate Change Committee (NCCC).

We then explained the roles of the various support and management players.

With the formal matters done, I suggested that the SBT give Harald the role of Reviewer, Critic and Suggester. I referred to Shakespeare's clown, who has the freedom to criticise the king. We needed someone above the fray who could ask the tough questions in matters of content. I was keen that Harald receive a mandate to make any input he chose at any time; to critique any team member's inputs, and generally to be 'untouchable'. Everyone liked the idea, and it was to prove invaluable over the course of the meetings. Our resident expert was free to push the technical agenda and he would do so almost all the time.

We then turned to the next and critical component of the process: the research element. Harald explained how this would work.

Research groups

The SBT would have access to four research teams, and would in effect *commission* the teams with SBT *agreed* input data. This simple formula was key to the methodology of the LTMS. What then came out of the research would drive the packaging of the Scenarios.

In many ways the Energy Research Team would be the main group, given that four-fifths of South Africa's emissions are from this sector. The team would use an internationally accepted and known modelling programme: MARKAL. This became the main tool for building the reference case: Scenario A. The Non-Energy Emissions Research Group would use Excel and other tools, and these would be combined with Group 1. Group 3, the Macro-Economic Implications Group would then study the results and model the economy-wide impacts. We could then see who the winners and losers were in particular mitigation cases.

'The ERC, the main service provider, had a good grounding in development. So it was not just a number-crunching modeller. This sensitivity to our development aims as a country was very important. The LTMS was grounded on development issues; it was informed by the broader social challenges facing South Africa. In addition the local service providers, being local, were very committed, and often they went beyond their brief and worked with a sense of common cause.'
Richard Worthington, formerly Earthlife Africa

The Impacts Group would provide us with a valuable insight into the Climate Change impacts South Africa could likely expect.

Rules of the game

The management team had agreed that we should first lay out all rules of the game, and then set out the structure for the Scenario planning itself. The first part was easy: with perhaps one exception – confidentiality. We were worried that given the need for the exposure of proprietary facts from SBT members about their sectors, members or organisations, we would encounter confidentiality problems and an unwillingness to part with information. For example: we would need to know what Sasol's expansion plans were going to be – a matter immensely confidential to that organisation.

Our solution was to apply Chatham House Rule, which is very well known, and is as follows:

> When a meeting, or part thereof, is held under the Chatham House Rule, participants are free to use the information received, but neither the identity nor the affiliation of the speaker(s), nor that wof any other participant, may be revealed.[30]

In the end, the question of confidentiality did not bedevil the process in the way that I feared it might – a testimony to the degree of willingness of the parties to the SBT to reveal their plans and even their associated costs. It does mean that the inputs of individuals are confidential, and naturally that extends to this book as well.

Status of work

At many points during the LTMS I had to remind everyone what the status of the work really was. This was evidence of a latent fear of the SBT members, namely that their work would in some way tie us to a path that was unrealistic, and perhaps that their individual reputations would suffer. For the NGO participants this was as pertinent as for the private sector players. I had to repeat: our work was purely speculative. We were building Scenarios. We would

> 'At various points during the process the debate re-surfaced: "What are we doing this process for? What will happen to the results?" We had to be reminded from time to time that we were creating Scenarios, this helped our mindset to think: "How will things play out? What type of actions can we put in place? Where will we be in 2050?" The LTMS was not a tool that solves all our problems but more a way of seeing what futures could play themselves out.'
> *Herman van der Walt, Fred Goede, Sasol*

30 www.chathamhouse.org.uk.

present an array of options to South Africa. We were not making decisions. Our work would also be subject to a degree of further consultation (with sector leaders and senior directors in government) before it went to Cabinet. Understandably, there was always underlying nervousness about this.

Definition of main terms

We decided that the SBT should use common terms, and the following were discussed and agreed upon.

- **Assumptions:** primary drivers for each Scenario would include assumptions and uncertainties;
- **Actions**: individual mitigation actions we may take as a country;
- **Action packages**: Agreed combinations of actions, with GHG emissions trajectories; and
- **Scenarios**: future 'stories', or paths, each populated with assumptions, action packages, sub-scenarios, data, including costs, benefits, trajectories.

In hindsight we could have done better in drawing a terminological distinction between the evidence-based trajectory Scenarios, the three Options, and the Scenarios we produced for future contexts. Terminological issues can haunt such meetings and constantly resurface. This one did bedevil the process somewhat.

Harald and I set up a Scenario framework for the inductive and deductive Scenarios and Options that we would eventually be building. We wanted everyone to get where we intended to go quickly. Our structure was a really early version, to change a lot later:

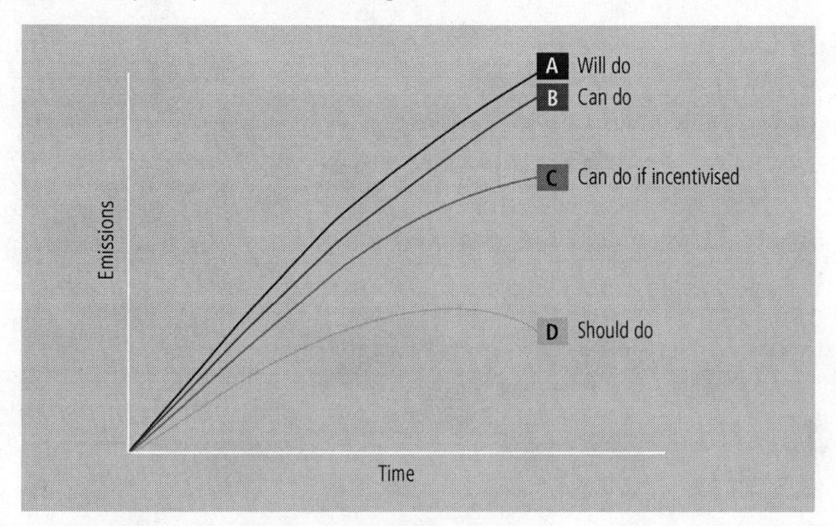

In the framework, Scenario D would be the Utopian Scenario, laying out what South Africa 'should do' if it had unlimited resources and was committed fully to the requirements of an international agreement aligned with science. This Scenario would serve merely as one of the two bookends, serving thus as an 'envelope Scenario'.

Scenario A is what South Africa will do. This one would be modelled and fully based on evidence. This evidence would reveal the emissions trajectory of South Africa up to 2050. This would be the other envelope Scenario.

Scenarios B and C could then be built as Options and would illustrate degrees of effort, and their costs. Hence the idea of effort we could afford as a country and effort with which we would need help. These Options would be built by 'stressing' the A Scenario. This means simply taking out elements of the A Scenario that have high emissions and replacing these with other technologies. In this way the Options would also be totally evidence based.

We agreed in the SBT to build Scenarios D and A first. Scenario A would be built by agreeing all inputs and assumptions. From the start it was clear that nothing would be entered into the dataset unless agreed. Consensus was key.

The *cost* of the packages of action within Scenario A (and hence the Options) was also a key piece of information. After all, it was only through an accurate determination of this cost that we would eventually be able to separate the Scenarios according to our classification (which draws a type of affordability line). This cost would be provided in the modelling process. Cost alone was important, but we also needed more: we needed to test the Scenarios for their socio-economic impact. We needed to know how much actions would cost, and what their economy-wide impacts and co-benefits would be.

We presented this framework proposal on PowerPoint for SBT approval. It was supported. So far, it was going rather well!

Of course no-one knew how big the gaps between the Scenarios would be. At this point I bet all of the participants were all getting curious. I certainly was and this curiosity was to prove a wonderful driver. We had a giant puzzle to piece together, without the picture on the box. So far we had only found the four corner pieces.

Background to the Scenarios: the South African context

I was still fascinated by the possibility of including in SBT 1 a context setting Scenarios exercise (in the classic way, building alternative non-evidence based stories). I believed we needed to 'open the minds' of the SBT members right from the start, and get them to go beyond merely ratifying a pre-determined Scenario structure. I talked through this with Harald, and we agreed that after presenting the structure (and if the team was happy with it), we should take them through an exercise to develop some alternative contexts for South Africa. This would shape the link between emissions reduction and development, which would in any event be uppermost in everyone's minds.

The context setting exercise was aimed at encouraging the group to think together about the problem of mitigation options. I was most interested in the outcome of this session and to see how the members would consider the issues ahead of us. In some ways the session was a chance for the group to 'stretch before the exercise', to limber up mentally.

In addition the exercise got us straight into the heart of the debate: how should we balance emissions reduction and the development imperative? How would we overcome the prevailing view that mitigating emissions was a threat to development?

We explained that we would do a basic quadrant-based Scenario play. After some debate the group agreed the 'Givens'. These are the main contextual parameters for the quadrant (see Chapter 2). Our givens were *growth* and *emissions*. A warm debate broke out on the meaning and impact of these two parameters. The opposing forces were laid out.

- To develop or not to develop?
- To mitigate or not to mitigate?
- Does development mean increased emissions?
- Can we develop without increasing emissions?

The group debated the meanings of 'growth' and 'emissions'. Was growth GDP only, or some other package of measurements? Was it sustainable growth or something else? Was negative growth to be included? How would growth targets be seen?

The group eventually agreed that the focus of this dimension should be growth, which is a narrower concept than development. This was *not* to say that other dimensions of development were not significant. It meant that the key tension was between putting effort into mitigation and achieving economic growth. This is what it looks like as a Scenario play:

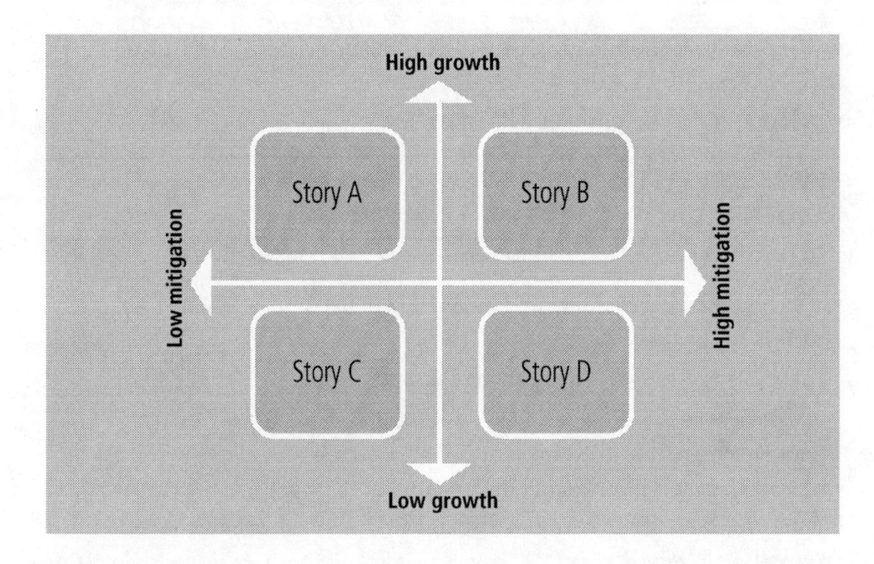

The four quadrants give us four possible background stories for South Africa. We did a quick Scenario play, not by any means a deep exercise. The group agreed to 'set' these four stories by listing assumptions and conditions, or headlines, which would characterise them. Some of the assumptions that were generated were, by way of example:

- Climate science is no longer debated;
- Energy demand increases around the world;
- There will be increased access to energy;
- Energy security is high on the global agenda;
- Energy prices will continue to escalate;
- Fossil fuels will continue to be a major primary energy resource;
- We will see oil and probably gas consumption peak soon;
- The population will continue to rise;
- Climate Change impacts will continue to manifest and become more intense;
- Population movements will occur as an adaptation measure;
- Emerging developing countries' emissions will overtake developed country emissions;
- The biggest emitting countries will have some sort of targets over the next 20 years;
- There will be voluntary actions for South Africa and others, however;
- Adaptation measures will have to be taken;
- Water is going to be a big problem;
- There will be a price for carbon;

- Generally economic growth will continue to be positive (no-one predicted the 2008 crash);
- Geopolitical realignment will take place;
- There will be global stability generally, with pockets of unrest especially with regards to resources;
- There will be a movement of manufacturing industry to developing countries;
- The roles of government and business will become nebulous (especially for big multinationals);
- African economies will start to pick up;
- The African continent will start to put pressure on South Africa to 'do something' about emissions; and
- South Africa will probably get an emissions target as a big emitter.

I think this was the first time a list of this kind had been generated by a group of such mixed views. And yet here were South Africans seeing the future in a remarkably similar light.

I broke the SBT up into four groups, and each was given the task to tell one of the four stories. The following was asked: 'In order to sustain your background story, what would have to prevail?' This is what emerged:

Story A is a default picture. South Africa grows and develops out of poverty, but relies on its large fossil resources to achieve this. It manages to avoid carbon constraint, to keep its energy prices low, and to focus on growth and poverty alleviation by all means. It does so even if it has to stonewall international pressure to mitigate. But this means that internally, political will is all but absent and democracy may be failing. There are many negative impacts, such as local environmental degradation, but growth is the only objective. South Africa is once again internationally isolated. This time it would suffer isolation for the sake of growth, and not ideology.

Story B portrays a South Africa where high growth has still been achieved, but that this growth is decoupled from high emissions. This is a South Africa with high energy prices, a diversified energy economy, and futuristic technologies. Democracy is robust and political will is strong. Internationally the country occupies the main stage, on the mound of the moral high ground. This is a 'breakthrough' view of a future South Africa.

Story C is one of deteriorating growth together with low mitigation effort. Energy costs are high, capacity is failing South Africa, and coal is being depleted. South Africa either fails to cooperate with international mitigation efforts, or is opportunistic as international negotiations fail. In the latter case, growing climate impacts in South Africa complete a miserable picture. Political processes are in a shambles and growth is slowing all the time.

Story D has two possible faces: either deteriorating growth is causing lower emissions, or efforts to lower emissions are having negative impacts on the economy resulting in deteriorating growth. Both possibilities would cause a crisis and an economy in collapse. The second version of the story, especially, raises a red flag: taking on the mitigation challenge and failing, thus overstressing the economy, would result in a bleak future for South Africa.

Of course, like all classic Scenarios, one is drawn to the 'success' story. The mind quickly rejects the other three. So what is the value of this? What did the SBT players learn? The group started to look at the stories as emissions and growth trajectories:

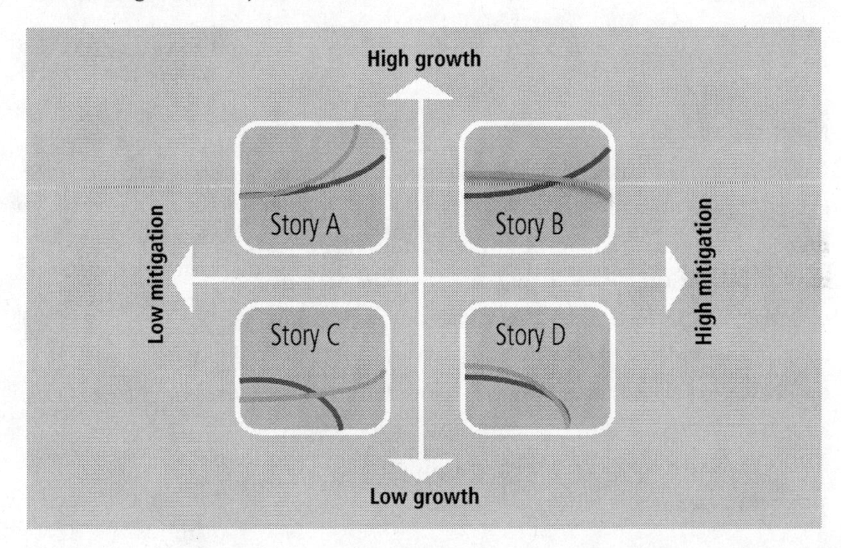

One thing we had learnt was that we didn't want a South Africa out of step with the rest of the world – a new environmental pariah. Neither did we want a South Africa that was poor, underdeveloped and a shambles. We wanted the South Africa of the future: growing and organised. But we needed our emissions to be much, much lower.

The SBT agreed at this point to revisit the 'negative' Scenario contexts at a later point, and to focus on the 'positive' Scenario. We felt that there was no point in presenting the 'failure' based Scenarios to South Africa. No one wanted story C or D anyway, so what was the point? Stories A and B were more interesting and provided scope for further analysis.

These two stories matched the framework we had set. 'Business as Usual' was Story A, and Story B was of course the decoupling of growth from emissions – the favoured route. But although we all wanted this outcome, I could almost sense the range of internal responses in the room: the NGOs would favour story B, but they had no idea of how to actually achieve it; the coal and fuel group would assume that this option was impossible, and so would remain fixed on story A; and the policy-makers would be wondering what to do next.

If anything, the setting of these four Context Scenarios had achieved the purpose of illustrating to all that we were in unknown territory.

It was a good way to end SBT 1, on a profound note of uncertainty.

'Right from the start we were encouraged to think ambitiously. We were all coming at this from different directions, but putting all the issues on the table up front helped. For example the debate environment/development was put on the table right up front, but in a structure that avoided the "value debate".'
Imraan Patel, DST

'External factors also helped: the Stern Review overlapped with the LTMS, and introduced a language that business understood; there was AR4, which put the science beyond doubt; there was Davos, and other events and initiatives. All these helped the LTMS to become a very credible process.'
Shaun Vorster, former Adviser to the Minister

Next steps

I remember heaving a sigh of relief as the meeting closed. It was all working well so far. The SBT members had agreed that the building of the Scenarios would be through the listing of possible actions, which are then assessed by the research teams against cost (in the widest sense) and emissions impact. They had had a go at the mitigation/development debate, and had loosened up, relaxing into the process. There had been no large procedural stoppages. They all *wanted* to do this work.

Chapter 4

The picture emerges

Chronology of the LTMS study at SBT level

SBT 2: 12 September 2006

SBT 3: 29 November 2006

SBT 4: 30 and 31 May 2007

From SBT 2 onwards we started to build the picture, piece by piece. The approach of 'component agreement' was to be one of the most important parts of the collaboration in the SBT process, with very interesting results from the perspective of outcome: the emergent nature of the process was defined by the fact that no-one actually knew what the overall results would look like, until they were actually achieved. Groups, depending on their particular interests, often do not like or support certain positions if presented upfront. Such groups could either totally withdraw agreement or pull out of the process, and the overall failure of the process would be assured. Proceeding component by component meant that if each input, assumption and result of each component was first agreed, the resulting end position would be hard to refute. Some smaller component issues did become deal breakers, but these could be packed away through various methods: we could, for example, simply omit a particular assumption – overall figures would be slightly affected, or accuracy compromised, but no matter – in this way we could deflect a full scale conflict on the issue.

We would find that this approach would be very successful in the SBT, as parties from all sides of the spectrum worked together on each of the

'Breaking the problem down into small components doesn't avoid the value discussion but does contextualise it. The key issue here is grounding the discussion. A context free space without any reference points, without some form of tie-breaker, simply results in a lot of hand waving. An assessment is a process that is about accumulating information and sorting it, and doing so in a consultative way. That information is then put in such a way that it is helpful in building policy.'
Bob Scholes, CSIR

components, with high degrees of trust developing as they struggled through the figures and statistics. There was a focus on details, not general positions or values.

'You are giving concepts their own domains, their own magisteriums.'
Bob Scholes, CSIR

We'd had a good start at SBT 1; in SBT 2 we moved on in earnest with the SBT process. The planned approach was essentially as follows:

The SBT itself would be used as a plenary to ratify work done by smaller stakeholder groups and the research teams, as well as entering its own agreements into the research work. In broad terms, the SBT would approve all inputs and assumptions (such as, say, the discount rate to be used, or a projection of the oil price). These would then be processed by the research modelling teams. The SBT would approve all outputs.

This sounds simple in theory, but in practice what this meant is that the SBT would have to reach consensus on every input. This would be difficult with many of the issues, given the wide range of interests and value positions of the SBT members. As an example: the SBT would have to agree all cost figures, including the technology learning applicable to certain technologies, where costs increase or decrease over time as technology either becomes more or less expensive due to improvements and/or scale. This can be very controversial: just try getting one view on the future costs of nuclear power. I had to resolve, deflect or avoid conflicts during the process. However, we agreed up front that no result would pass by the SBT as a group: it was imperative that every result in the end had an 'approved by SBT' stamp on it. This is a foundation element of the LTMS: the Scenarios were the work of all.

'The interesting thing about the SBT was that it developed a level of comfort and consensus in such a way that people could take positions which they could not necessarily have taken within their own organisations.'
Kilebogile Maroka, DEA

The volume of work for a project like this is very large, and the SBT in plenary cannot deal with the preparation work: smaller stakeholder groups did this. These groups would basically set up the work in their sectors. Hence, for example, in the agricultural sector, a smaller group of technical experts was formed to meet and determine the extent of the mitigation actions that could be achieved in that sector.

SBT 2 and 3

The second SBT took place in Pretoria on 12 September 2006.

At the close of SBT 1 participants had been given some homework. They had been requested to identify mitigation actions in their own areas of expertise and sectors. In addition they had been asked to list key drivers (for example, one of the most significant of these drivers would be the agreed GDP trajectory for the country, right up to the horizon year, 2050).

We repeated that the five Scenarios to be developed were:

1. Current Development Trend (CDT);
2. Business as Usual (BAU);
3. Can do;
4. Could do; and
5. Required by Science (RBS).

At this stage these Scenario names, and their meanings, were essentially provisional, and as will be seen, some were to be abandoned at a later point. For the moment they served only as a notional framework.

'This is probably the first time renewables, nuclear power, fiscal instruments and other interventions have shared a common platform in one integrated study. One of the agreements was that everybody could see from the numbers that not one of these technologies alone would be the solution.'
Kevin Nassiep, CEF

The somewhat clunky distinction between CDT and BAU persisted right to the end of the SBT process, being essentially the difference between a totally unconstrained approach, and existing policy.

The SBT would first build the two envelope Scenarios. Scenario A would be the 'reference' from which the Options were extrapolated, and Scenario B would provide a contrasting picture.

For the sake of coherence, I will depart somewhat from the chronology of the process, and focus instead on how these two Scenarios were constructed.

The base year, and the horizon year

The determination of the starting point and end point of the Scenarios was naturally very important. The first is a technical challenge as it relates to the agreed absolute emissions in a year. In fact, such a figure was not readily at hand! South Africa did not have accurate, up to date emissions figures. There was no inventory. We had to work out a way of getting the right figure for a starting date. In the end the year 2003 was agreed, and we set about agreeing the level of emissions as at 1 January of that year. This meant looking back to the best figures we had from a 1994 inventory, improving and reviewing those figures, and using known growth figures for the main sectors to work out the total absolute emissions for South Africa at our agreed start date. In the end this was 443 million tonnes of CO_2 equivalent in greenhouse gas emissions.

The end date was less complex. The most commonly referred to dates in IPCC literature for future reference points are 2025 and 2050. We wanted to look far ahead, so 2050 seemed logical.

Building Scenario 5

This Scenario was built very quickly, using the normative approach (where you want to be in end defines the method of getting there). It was agreed that the Scenario should be called 'Required by Science' (RBS). The team agreed that this Scenario would see South Africa 'doing what science requires' to help achieve the 2°C global mean temperature stabilisation target. It was agreed that this meant a reduction of South Africa's emissions to between 30% and 40% of emissions levels in the base year. This is derived from AR4,[31] which indicates the need for a 60–80% reduction in world emissions in order to achieve stabilisation. In the RBS Scenario South Africa would have reduced its emissions by half of the required total, given its status as a developing country. This 'developing country discount' was agreed by the SBT. It is, or course, a very important assumption!

RBS was produced in one part of the SBT 2 session and was the fastest Scenario to be conceived by the SBT. Its purpose, at that point, was merely to 'frame' the Scenario set – to show what South Africa would have to do if all conditions enabled it to do this.

> 'Why the LTMS approach worked was because it was simple without a proliferation of different Scenarios all based on different assumptions.'
> *Bob Scholes, CSIR*

The SBT made note of the fact that implementing the Scenario would probably wreck the South African economy. Utopia, even with some slack, was just an aspiration. That was certainly the prevailing view in the room. Everyone in the room *assumed* that RBS would be destructive, but for the sake of completeness, the RBS was posted up and, for the moment, left in the 'park' position.

This outlandish Scenario was later destined to become hugely significant. It was going to be the surprise of the LTMS.

Harald and I agreed that presenting the material in graphic form was very important, and I started playing with various PowerPoint slides. Using animation techniques, we started building the graphic presentation of this work – it was to be invaluable later. It was fun too, drawing complex curves and turning them into 'art'! We weren't alone: I noticed Peter

---

[31] Intergovernmental Panel on Climate Change Fourth Assessment Report: www.ipcc.ch.

Lukey from the Department scribbling a lot, and he eventually produced the LTMS logo to great fanfare. These time-consuming efforts were not purely whimsical though, as they helped to 'brand' the LMTS from the bottom up, and to present the work visually in an arresting way.

The resulting artwork illustrates Scenario 5. One can see from the graph that the emissions curve is a theoretical curve from the base-year to 2050, declining in a 'cloud' towards a range of –30% to –40% of the emissions at the base year.

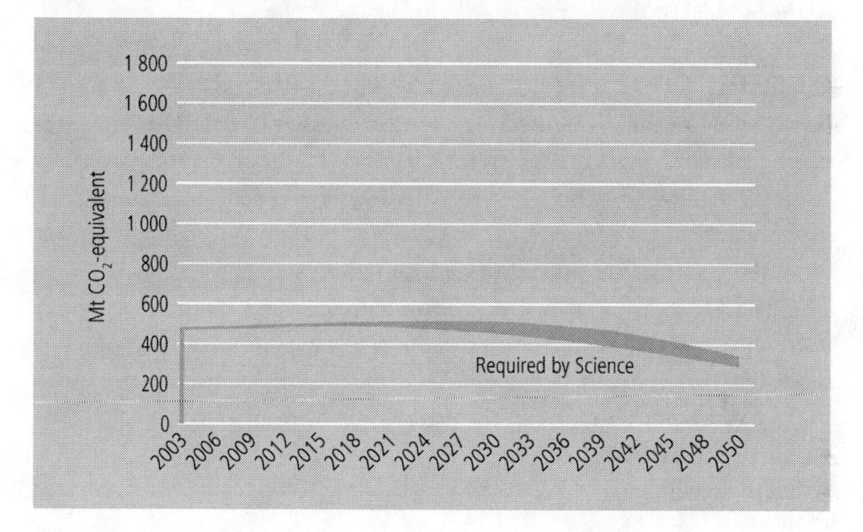

We moved on to Scenario A.

Energy emissions: the big job in Scenario A

Given that most of South Africa's emissions are attributable to the energy sector, the Energy Research team would be doing the lion's share of the work on Scenario A. The team would be using a customised version of an existing MARKAL modelling programme. MARKAL would be used to develop the driver Scenario, which we also called the 'base case'. I learnt that these types of models basically do the following:

The SBT would supply the model with a brace of inputs and assumptions, such as the period for which the model would operate (from 2000 to 2050), the growing (and declining) population of South Africa over the period, and the level of GDP desired. The cost of various (known) energy options (e.g. coal-fired electricity in its various forms) would be provided, thus giving the model a number of options to 'fire up' the size of economy presented by South Africa over the stipulated period.

> 'During the SBT phase we had to maintain scientific objectivity and purity in the work.'
> *Kevin Nassiep, CEF*

Once all the inputs and assumptions had been agreed, the computer would calculate the cheapest options for energising the economy, year on year, and would constantly add the growing emissions of the economy. If, for example, wind energy became cheaper than coal in 2035, it would automatically start including this zero emissions form of power production, according to the rate at which the country could install wind capacity, another input that would be provided to the model.

Scenario 1, after some debate, became known as 'Growth without Constraints' (GWC)[32] (with 'Current Development Trends' being dropped from the top position and used to reflect Business as Usual [BAU] instead, given that BAU would reflect compliance with policy). GWC was to become the 'driver' Scenario, given that all the agreed inputs and assumptions drove that Scenario, with Scenarios 2, 3 and 4 essentially being stressed versions of Scenario 1, or a set of options to Scenario 1.

The SBT started agreeing on the assumptions and inputs that would make up the terms of reference for the work by the research groups in modelling this Scenario.

Here are some examples of the drivers that were agreed on:

- Discount rate;
- GDP growth rate;
- Population;
- Reserve margin;
- Base year;
- Prices (currency and year);
- Exchange rate;
- Domestic coal price;
- Oil and gas prices; and
- Technology learning rates (a working group was formed to deal with this difficult component).

Building the Scenario required work spread over three SBT meetings, and in between, given the multiple bits of data agreed at the SBTs, the Energy Research team laboured away at the University in Cape Town. I spent

---

32 This Scenario was actually constrained to some extent, leading the SBT to moot the suggestion that it should be called Growth without Climate Consideration, which while more accurate was a bit to clumsy; for details of the Scenario, see Harald Winkler's book on the LTMS.

many hours there, observing their efforts and enjoying the interaction with this warm, talented and immensely dedicated group. The work took months.

Eventually they could run the model, which was immensely complex. The team pressed the 'enter' button on the MARKAL model (more or less!), and a new emissions trajectory emerged. It was astonishing. It was eventually ready for presentation to SBT 4, with one more working SBT in between.

SBT 3 and 4

SBT 3 had taken place on 29 November 2006, at Emperor's Palace in Kempton Park. At this meeting we had presented the first draft of the text that eventually would become the report of the SBT: the so-called 'Technical Report'. The document was already 75 pages, and filled with dense lists of figures.

Here is an example of the types of issues that were traversed, this time in the batch of technology learning assumptions:

- The following text was added: 'The rate of the doubling is based on the historical growth rate reflected in Table 4. Technologies will grow until they reach a maximum global capacity. The SBT agreed that where the research teams could not find maximum global potentials in the literature, they would assume an estimate.'
- Two additional columns were added to the table, which the researchers will populate: 'Range of learning rates in literature' and 'Maximum level this technology can reach globally'.
- The learning rate for Carbon Capture and Storage will also be included in the Table.
- It was agreed that the researchers would consult with the DPE to confirm what has been discussed.
- The issue of skills shortage was also noted, although it may equally constrain all technologies. This may be a step to consider after the LTMS process.
- All figures in this section will be circulated to the group shortly for a quick sign-off process. All comments will also be circulated.
- The section on the PBMR was accepted, with the addition of the following text: 'On the PBMR costs, it was accepted that a range of costs need to be considered and therefore a Scenario should also look at other costs based on the closest equivalent technology.'

Not easy stuff this.

SBT 3 was gruelling by process standards. My 'single text' approach threatened to derail a few times, due to the sheer impact of the painfully slow process. Morale flagged. I remember looking around at the bent heads and wondering if they were still committed to this.

The LTMS was becoming a forest of data.

The fourth SBT took place in May 2007 in Centurion. We presented the GWC Scenario to the SBT in plenary. The result was a shock to everyone. We all knew that our emissions would rise in the unconstrained Scenario, but by this much? South Africa's emissions would almost quadruple over the period. This is what the graph looked like:

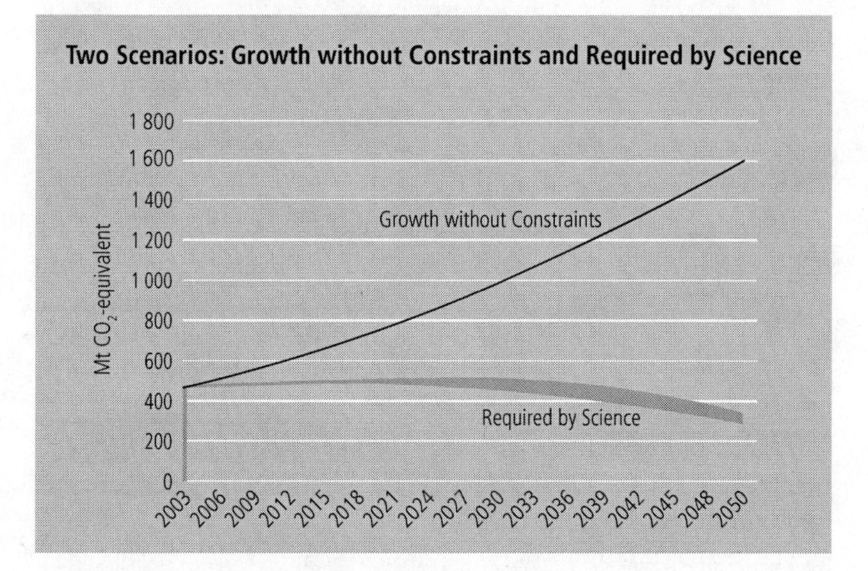

A yawning gap had appeared between what *would* happen and *what* should happen. And the gap itself was a lot larger than South Africa's current emissions.

This moment had a strong effect on SBT participants. Not only did it show how much our emissions would increase, and how that might affect our position in the international negotiations, but it also revealed just how much energy would be needed to fuel our growing economy – it hinted at what was to come: an energy crisis for South Africa.

All the team members in the room knew that this was a trajectory they had built, so there was no dispute, no 'that can't be true' argument. There were some who wanted to re-interrogate the figures, and some members did point out that certain inputs and assumptions could be seen as compromised, but we all agreed that these flaws would make very little difference to the net result that had emerged. What had emerged, whether it was a fourfold or threefold increase, was that South Africa's natural development path was hopelessly coupled to a massive rise in its emissions.

'We did not have a deep enough questioning of the validity of all the numbers in the base case. But I agree that the shock of the fourfold increase was profound. I think the scale was so important: we may have disagreed with some of the figures, but it would still have illustrated the big differences between where we would be going and where we should go.'
Imraan Patel, DST

The component approach had worked, and the results were there for all to see. There was immediately a certain unity in the room – we had a problem. We turned to our current policy package, which we could then with some ease illustrate against the base case. But when we 'extrapolated' the current renewables and energy efficiency policies, a second lesson was learnt. South Africa's existing energy mix policy package hardly made a dent in the extrapolated emissions.

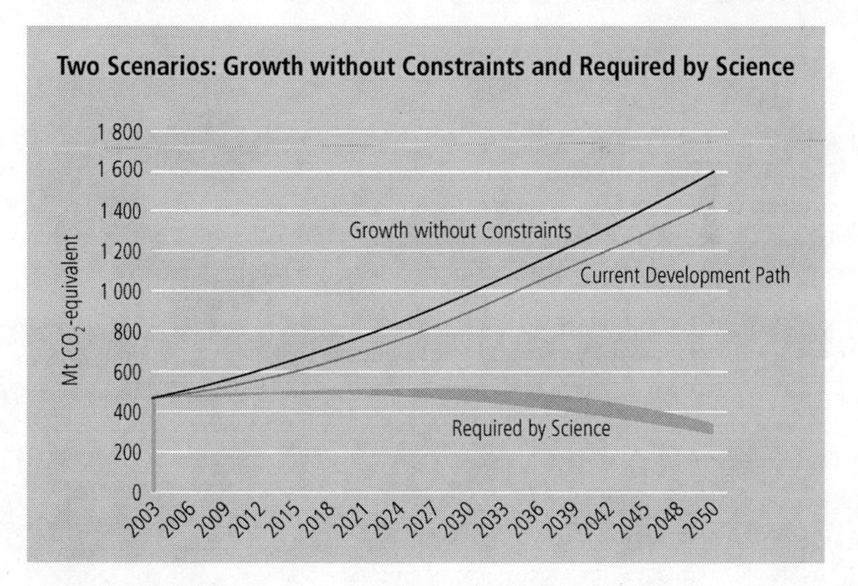

And to make it worse, as Harald pointed out, CDP would still require significant mitigation effort and progress by 2007 was not exactly blistering.

This raised a fresh challenge: how could one close this gap? What actions were needed, and what would be the cost? It was as if the climbing curve had set the challenge.

'In the LTMS we didn't follow a standard Scenario process. The surprise in the LTMS, for me, was that we very rapidly converged on two Scenarios; the first was the Business as Usual Scenario which was in a sense quite self evident and for that reason did not have to prove itself, given that it was based on documents on the table, often being government stated policy or figures already accepted by government. The second was the Required by Science Scenario which was also of external origin, being the IPCC work fine-tuned into the South African context. These two Scenarios emerged organically. This greatly simplified things. You could then run these two Scenarios through the process of assessment. Now this violates a lot of the rules of Scenario work (the rules say never have two Scenarios!). Technically Scenarios should be equi-probable: there should be no a priori reason for choosing one above the other. In breaking those rules and coming up with a really stark choice, the LTMS set two very clear alternatives, both at first equally justifiable, and it was very clear which one would be defensible.'
Bob Scholes, CSIR

The SBT now faced a new and immense challenge: finding out what it would take to close the gap.

Exploring the actions

Our strategy was to look at every possible emissions-reducing action that could be taken in any sector of the economy. We restricted ourselves to known actions, those that could be costed and their impact on emissions calculated. We wanted hard evidence.

In SBT 3 participants had been broken up into three groups:

1. Energy generation;
2. Energy users; and
3. Non-energy emissions.

Each was requested to create lists of mitigation actions for that sector. This process was essentially an exercise in making lists of mitigation actions, and assigning to these lists agreed figures which showed the degree of penetration of the action into the economy (such as the introduction of electric vehicles, and the timing of that penetration).

To illustrate this, here is an excerpt from the list for transport:

'The important thing was that before the LTMS the issues of Climate Change were being articulated by the NGOs, who probably believed that Eskom and other players were not prepared to act on Climate Change. In fact Eskom had done extensive thinking and had developed its climate strategy before the LTMS started. But when we spoke about the Growth without Constraints Scenario, I could see that everyone, NGOs and the emitters, realised that we were on the same page in terms of wanting to act on Climate Change.'
Mandy Rambharos, Eskom

Action	Penetration	Timing
High efficiency engines	30% new vehicles	2010
Hybrid vehicles	10%	2025
Use of public transport (trains, busses – metro) e.g. congestion taxes, virtual working	80%	2015
Transport demand management mechanism, e.g. park and ride	80%	2015

Where these actions were cheaper than more polluting actions, MARKAL had already selected them for the GWC Scenario. Now we had extended these actions. It was in two areas that the larger mitigation results would be felt: energy generation and mobility. Within the first, two forms of electricity generation stood out.

Nuclear versus renewables

Given our reliance on coal, it stands to reason that any alternative form of electricity generation will have a big impact on emissions. But amongst experts there is a polarising adherence to either nuclear energy or renewable energy, and differences are deep, and often value-driven.

My job was to facilitate the agreements on inputs. Some of the arguments were of a highly technical nature, and were conducted in a good spirit. The discount rate discussion is an example of this. But others were fought over real differences of understanding and data and the ideological fence also divided some of the SBT members. This was especially so on the issue of nuclear power versus renewables and we had to make a decision on what to include in the model and what to exclude.

The dominant position of the civil society SBT members was against a reliance on nuclear power, for a number of reasons: cost, security, centralisation and so on. These are familiar and strong arguments, strongly held. On the other hand, Eskom's energy mix planning for South Africa, and indeed their own Climate Change Response Scenarios (which they had been working on at the time of the LTMS) relied heavily on a number of nuclear stations taking on the bulk of the base-load requirement for South Africa from around 2020 onwards. Some business interests supported this view. Industry needed base-load power, and nuclear seemed the only choice.

The differences between the players are sharply felt and worded, but we found consensus on the following basis: the LTMS study would be 'non-ideological', and would study *all* technologies in the mix. This was for some a reluctant consensus, and if ever the LTMS is cited in support of a nuclear build programme, the finger will be pointed at the decision made by us in the SBT. I think that we were right, however, to have studied everything. If we hadn't, many would have discredited our work. It is bound to still be attacked, but one should, after all, try to understand all realities.

Chapter 5

Getting there

'It became obvious that business-as-usual would put us out of business.'

Richard Worthington, SBT member

Chronology	
SBT 5:	21 and 22 August 2007
WG 5:	September 2007
The 'fairy godmother' meeting:	6 September 2007
Economy-wide WG:	3 October 2007
Final WG:	3 October, later that day
SBT 6:	24 October 2007

Building wedges of emissions reductions

The SBT's attempt to close the gap took some months. The number of actions we needed to explore was large, and the detail extensive. We needed to see how each of the actions would perform, in terms of mitigation, and what it would cost. We could then also combine actions into packages. Some would 'subsidise' others in this way.

In each of the sectors, we needed to ascertain what mitigation action that sector could reasonably, but with a degree of ambition, achieve. These actions required the small technical groups to work with the research teams and create a 'wedge' of emission reduction.

This concept of the wedge is important to understand. A useful description comes from the WRI:[33]

> A 2004 paper by two Princeton researchers, Stephen Pacala and Robert Socolow, demonstrated graphically how a suite of existing technological options could be used to reduce GHG levels to a level that is sufficient to avoid the dangerous effects of Climate Change.

[33] World Resources Institute: 'Scaling up', April 2007 www.wri.org.

> The paper illustrates this point by breaking the required emission reductions down into manageable (though still large) 'wedges', each provided by a different technology or set of technologies. The concept is based on the comparison of a business as usual (BAU) projection of GHG emissions into the future with the desired trajectory of stable global emissions through the year 2050. The triangle-shaped gap between the two lines [see figure below] can be divided into smaller portions, each of which represents a technology option ... This visual illustration conveys a simple, powerful idea: that, despite the scale of the problem, we have the potential to solve it if we can deploy today's technologies at sufficient scale. The concept's resonance is such that now barely a discussion of climate technology fails to mention the wedge concept. It is 'the iPod of climate policy analysis ... an understandable, attractive package that people can fill with their own content'.

The wedges are then presented like this:[34]

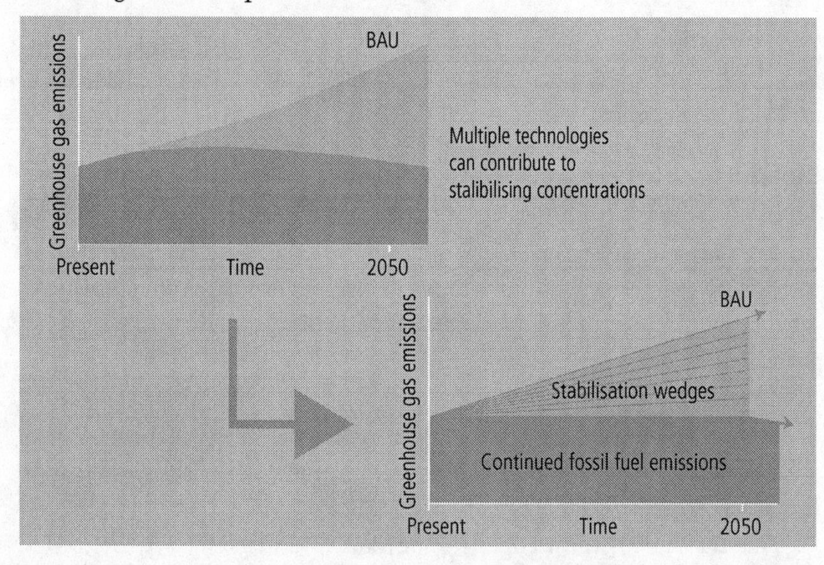

We worked constantly on these wedges outside the SBT; I stayed close to the research team and watched them 'live' with wedges. Dr Andrew Marquard, one of the seniors in Harald's team, became known as 'Doctor Wedge'. In the end many wedges were modelled, and some of these, (the main ones) were gathered by Andrew in a now much used illustration of the collection of work:

[34] Ibid, WRI paper.

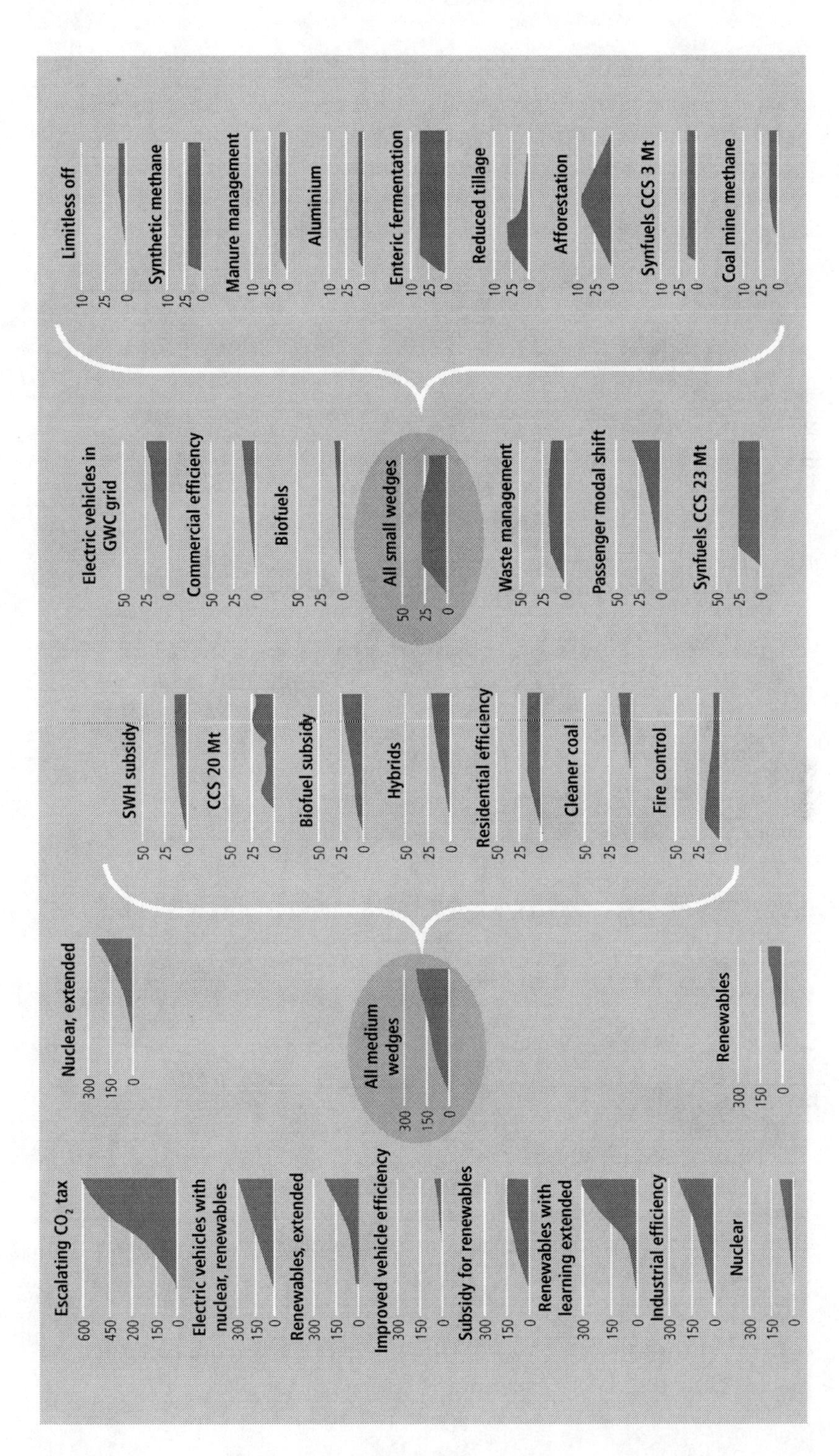
Escalating CO_2 tax
600
450
200
150
0
Electric vehicles with nuclear, renewables
Renewables, extended
Improved vehicle efficiency
Subsidy for renewables
Renewables with learning extended
Industrial efficiency
Nuclear
Nuclear, extended
All medium wedges
Renewables
SWH subsidy
CCS 20 Mt
Biofuel subsidy
Hybrids
Residential efficiency
Cleaner coal
Fire control
Electric vehicles in GWC grid
Commercial efficiency
Biofuels
All small wedges
Waste management
Passenger modal shift
Synfuels CCS 23 Mt
Limitless off
Synthetic methane
Manure management
Aluminium
Enteric fermentation
Reduced tillage
Afforestation
Synfuels CCS 3 Mt
Coal mine methane

At SBT 5 the group continued with the work of transacting the gap, in what was an iterative process of dialogue between the SBT and the research groups and between the intense effort of the sector working groups and the SBT itself.

Eventually we were in a position to build the first 'package of actions'. This would become the first of the three Options, named Start Now. Start Now was a package made up of aggressive energy efficiency measures

Industrial efficiency

300
150
0
–R34

together with some measures in the transport sector

Passenger modal shift

300
150
0
–R1 131

Improved vehicle efficiency

300
150
0
–R269

and finally nuclear and renewable energy, in equal measure.

Renewables

300
150
0
R52

Nuclear

300
150
0
R18

In each of these, the wedge is shown as well as the cost per ton of the emissions reduction. The first three actions are cost negative, which means that they save the economy money, relative to the base case (the unconstrained GWC Scenario). The second cost more than the base case. Together the cost is less than the base-case. Hence the package is more or less a no-brainer – one you'd better start now. Hence the name.

The amount of nuclear and renewable substitution was about 50%, or put another way, by 2050 we were replacing around half of the coal-to-energy economy with these technologies.

We entered the package into the computer programme, waiting to see how much of the gap we would cover. The result was not encouraging: it was about 40%. The cost was good, and emissions reductions were marked, but not even halfway.

The SBT asked the research teams to extend the package. The energy efficiency measures were at their most ambitious, and so the real extension was by replacing all coal-fired electricity with nuclear and renewables, again in equal measure. Some Carbon Capture was added, as were electric vehicles:

Industrial efficiency
300
150
0
–R34

Nuclear, extended
300
150
0
R20

Renewables, extended
300
150
0
R92

Synfuels CCS 23 Mt
300
150
0
R54

Electric vehicles in GWC grid
300
150
0
R607

So this group was a combination of the first plus an extension of energy generation to 50% nuclear and 50% renewable energy by 2050. This seemed breathtakingly ambitious. The entire coal-driven energy system of South Africa would have all-but vanished. The results would surely be more than sufficient. Once again the model was stressed. The result, however, was not enough.

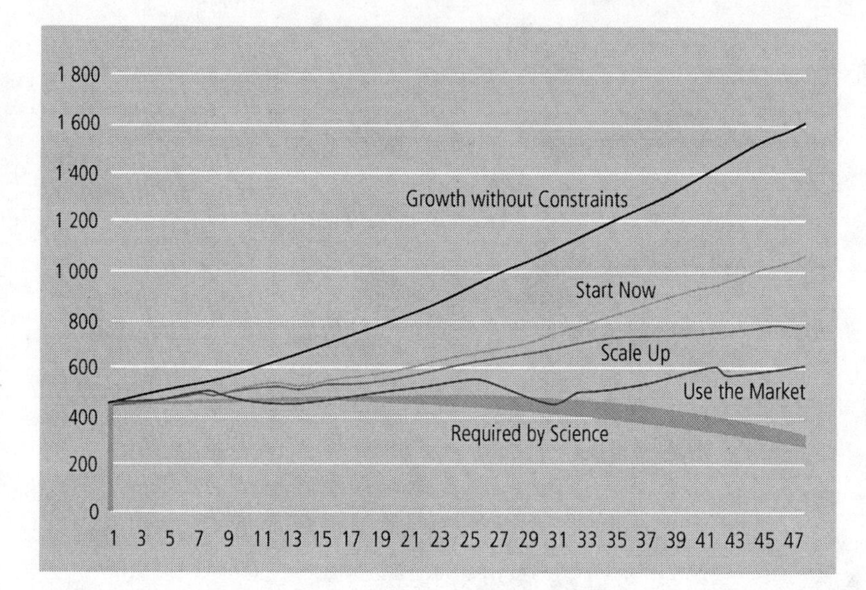

The black line shows the first of the Options, which we had called Start Now. The dark grey line shows the result of stripping all coal out of the economy, the Option we eventually called Scale Up. This Option still did not cross the gap, and essentially we had now run out of actions that would have big results. Now the options would start to cost the economy money.

We were able to add up the costs of the actions, and produce an 'abatement curve'. When one views the costs of each of these actions, the following illustrates how the actions progress from cost savings (reflected above as 'negative costs') to real cost interventions.

The combination of packages was yet another of the LTMS 'revelations': some mitigation options look very costly on their own, but in combination with others, the costs looked very favourable indeed. What we could do was not necessarily so crippling cost-wise. Or so it seemed.

> 'From the LTMS we could also see how technologies could compete economically in contributing to this solution.'
> *Kevin Nassiep, CEF*

However, the SBT wanted to try another route: using the market to drive emissions actions. This is of course distinct from mitigation actions, which are either simply 'executive actions', driven by policy or profit. Market driven actions are driven by taxes and incentives. The SBT wanted to try a package of these and see what the result would be. Given the price signals that these would send, the model would once again recalculate the best way to 'fire' the economy we had designed. We would then see the result on the emissions trajectory.

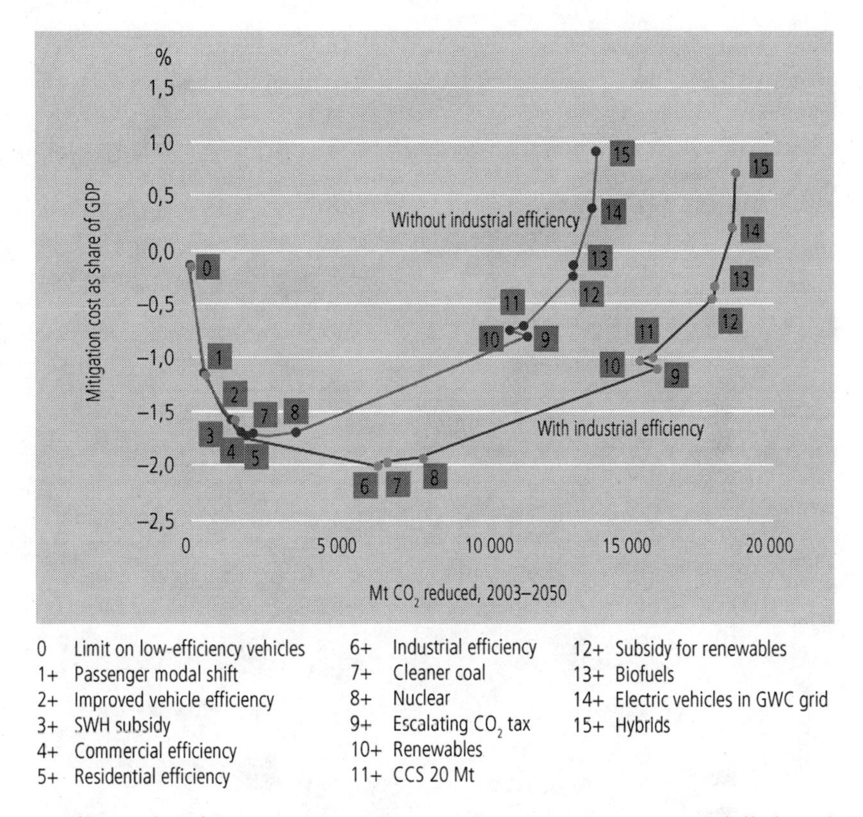

In the market driven Option, a progressive tax regime was modelled, and should be seen as a stand-alone alternative approach. It illustrates what would happen to emissions if such an economic instrument was introduced into the South African economy. With it were some subsidies:

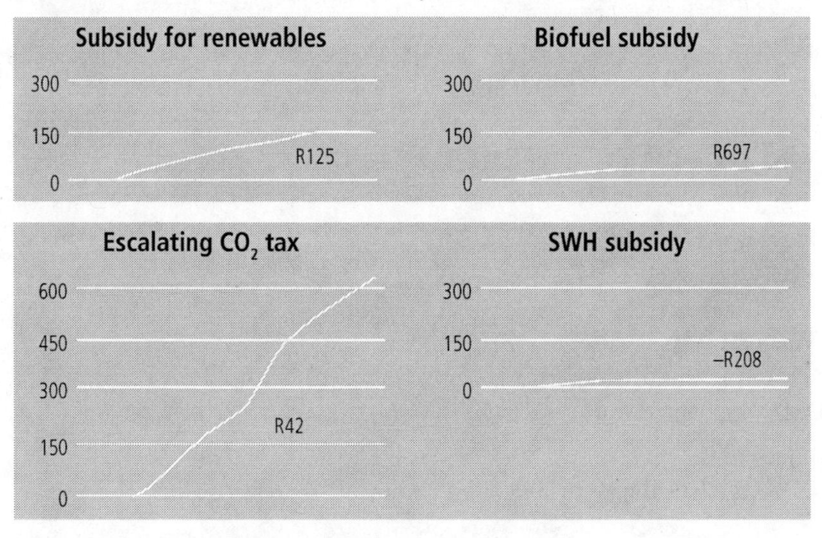

The carbon tax starts immediately at R100 (slowing emissions growth), grows to R250 by the mid-term (stabilising emissions), and then on to R750 by the end horizon.

The impact of the market package was encouraging, as can be seen in the jagged white line:

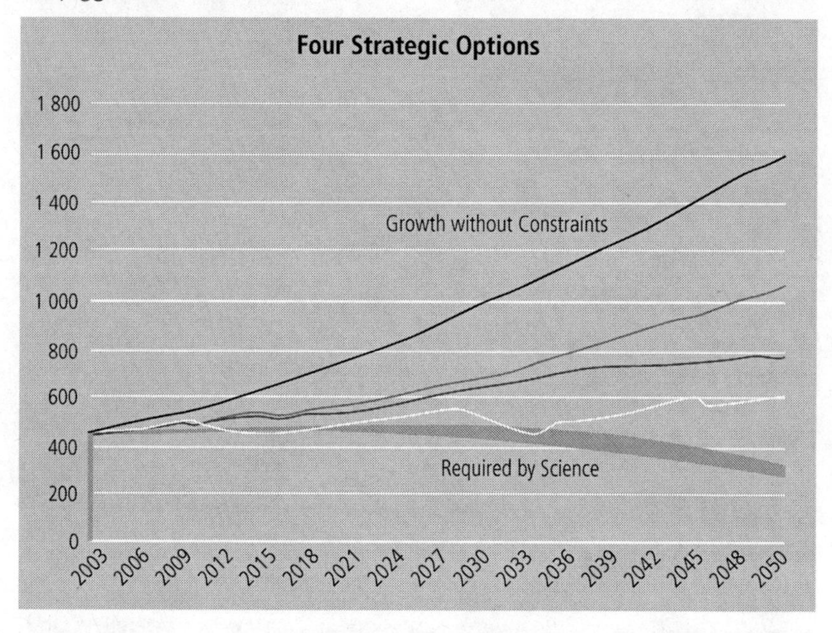

The package was controversial, but was adopted. It appears to work well in terms of emissions reductions until 2035. However, the gap is still there after that. We had not found a way to close it.

The SBT now had completed five trajectories, all of which were providing an emissions reduction picture, as well as a cost per ton of CO_2 reduction achieved. We had, however, run out of options. Nothing further would close the gap, it seemed. A few team members started to point out what was missing from our study: new technologies that would emerge over time; new energy and resource discoveries; new behavioural approaches; and a complete shift in the economy. These were worth talking about. These *had* to be explored, that was clear.

The last gap

There was still a gap between our most aggressive Scenario and the RBS Scenario. This was due, to some extent, to the fact that we had agreed to model only those technologies that we knew were viable and that we could cost, therefore the model ran out of cost-effective and known technologies in about 2030. It was apparent that by then we would need to have some new arrows in the quiver, but we had no idea as to what they would be. A hydrogen economy? Waterless washing machines? Cars that were power stations? How much was real and how much would simply be pegging our hopes on a science fiction rescue mission?

We were really in unknown territory beyond 2030 or thereabouts.

Technology too, was not everything. The base case represented a particular type of economy that we were growing and energising – a highly energy intensive economy. A far less energy-hungry economy that still performed for its people would have much lower emissions. The structure of the economy was really important. The behaviour of people in the economy was also important. Their patterns of consumption and other patterns of conduct could also have significant impacts on emissions. Finally, there were resources still undiscovered, or if known, still too difficult to exploit. Examples of these are the hydropower resources of the Congo, the tidal push of our coastal currents, and other energy finds, such as more offshore gas.

We needed to do some more thinking, but had no time, and less evidence to build a case. I proposed that we have a look, just a quick look, at those technologies that we had not modelled and that could have an impact on this gap. The group agreed. We needed a fairy godmother, after all.

The day of the 'fairy godmother' technologies

The Department of Science and Technology is housed in a sleek, new, green building overlooking the east of Pretoria. I prevailed on Imraan Patel of the Department, who hastily put together a group of technology experts.

We told the group which technologies had already been modelled, so that these could be excluded from the discussion.

I introduced the process for a brainstorming exercise in which the group could think up a list of 'fairy godmother' technologies: technologies of the future, that were tangible enough to be taken forward into the market, and known enough to ascertain their impact on emissions. I proposed that we first identify the technologies and then assess their likely impacts, time horizons, barriers and so forth.

First we generated a full list of possible technologies, and then we classified them against their mitigation potential and viability. A list emerged, which some time in the future may prove to be helpful. Within a day we had an interesting exploration into the wedges of the future, both large and small.[35]

[35] See Annexure 9.

Perhaps in the next LTMS, we will be able to explore these interventions further, and get a glimpse into their scope, cost and impact. We would look at:

1. The 'fairy godmother' technologies. Which would be the biggest and most cost effective? Which would produce jobs and help our people out of poverty? Which could go most swiftly from laboratory to market?
2. Newly identified resources, such a Hydropower outside of South Africa, and potential gas finds offshore and elsewhere.
3. People-orientated measures, with low or no cost to the economy. Examples include having two time zones for the country, or staggering traffic, school holidays, or work hours. The consumption patterns of people, and most significantly, their expectations of energy provision, their proclivity to large cars and energy hungry homes, would all come under this heading.
4. The transition to a low carbon economy has perhaps the greatest potential of all. We would need to consider business 'unusual' for South Africa: the growth of those businesses that yielded less carbon and more jobs.

SBT 5

It was, in a way, fitting that the fifth SBT should be held at the offices of the South African Reserve Bank in Pretoria, as the focus was increasingly being directed at financial and economic aspects of the LTMS study. It was August, almost exactly a year since we had started.

So much had changed. Harald had developed immensely as a leader, and I had held the process together. We had both enjoyed the planning immensely. The SBT team was a relaxed, affable group now, with high morale and bubbling with creativity. Tony Surridge, one of the team members, had become the team poet, writing astonishing couplets to the amusement of everyone. Industry members sat in relaxed debate with NGO members, and they viewed one another as individuals rather than interest groups. I saw myself as being constantly stressed, rather withdrawn and focused most of the time. Later Pascal showed me that good facilitation includes being more affable!

SBT 5 opened with a lengthy debate on the ultimate purpose of the LTMS study. It is interesting how this debate came up from time to time. This time SBT members, sensing that the work was coming to an end, were keen to know what would become of the outcomes of the study. The truth was that the outcome was uncertain at that stage: the reaction of government to the Scenarios would be something that would only emerge after the High Level phase, and there was no guarantee that the study would have any impact at all. I made the following point in the SBT:

> The process of Scenario building is a process that shapes people's decisions, as they may come to inescapable conclusions that influence the strategies of individual players. The LTMS must process the data in as accurate and informative a manner as possible and form stories that present something so new that it shapes decision making. It must also create space for discussions to take place on how individual players go forward.

The SBT now had one purpose: to achieve sign-off by the SBT of the final report. We were looking for the final consensus of all team members.

A remarkable consensus

I remember that on day two of SBT 5 we were wondering around in circles, and my facilitation had a rather aimless feeling about it. Everyone was in good spirits, but not really clear as to what to do next. We had two Scenarios and three Options made up of packages of Action. The gap was not crossed. So what next?

Then the discussion broke free into the space beyond the technical issues relating to the Scenarios and Options. We started to see what might be a workable total picture for South Africa.

We had a debate, for the first time, about the plausibility of the Scenarios. For the first time team members started thinking about the Scenarios as a group. And for the first time one of the team members said that a plausible Scenario could include a peak and decline of emissions, with a 'soft landing' in the plateau.

Another defining moment in the LTMS had arrived. This idea was worth exploring. Mandy Rambharos from Eskom made a suggestion: why not sketch the global picture – the possible future global pictures? And test our Scenarios to see which was the most realistic? Pierre Mukheibir took a different route, suggesting a SWOT analysis of our Scenarios. Others resisted Mandy's process suggestion, saying it was a lot of work and would take up valuable time. I was also worried about the time management issue, but was enthusiastic about the idea. Mandy persisted, saying that if we did not 'sketch up' the background to the Scenarios, the study would be effectively incomplete. Harald proposed a thirty-minute discussion. Everyone was pressed for time, but in the end Mandy's reasoning won the day.

I reminded everyone that the reality check was to be conducted *as if we were living in 2050.* We had to imagine ourselves living decades from now, and then assess which of our Scenarios would have performed best. This would mean imagining the world as it would be then, which of course is impossible. So we would imagine different possible worlds. We would, once again, be constructing alternative contexts for the future – yet more Scenarios!

Future contextual 'Scenarios': the external environment

This time it was barefoot Scenario work. We quickly split the SBT into five groups, agreeing that each would come up with a quick 'snapshot' of the world in 2050. This would take the form of newspaper headlines as they might appear in 2050.

We had little time. No quadrants were used, and each group was given the freedom to come up with whatever story of the future they felt was most probable. I walked from group to group and realised that they were actually having fun. After the tough sessions dealing with data, they could loosen up and play mentally with the future. With five pictures literally forming on flipcharts, I watched as the groups, representing such disparate South Africans, took a shot at the task. I marvelled at the way the people of this country, when given the opportunity, could muck in and get on with one another and the task at hand.

We asked the group facilitators to report back. It was as if this session had not been taken too seriously. The stories were, as a result, fascinating, as were their titles.

From Bling to Ubuntu

Ubuntu represents the concept of African togetherness and consensual decision-making for the common good. If you don't know what bling means, then ... This is how the story unfolds:

- Things get worse before they get better, following some sobering disasters
- UN reforms, moving to global democracy, and the African Diaspora mobilises
- Multilateralism wins the day
- The World Bank is transformed
- A Tobin tax is implemented and arms are taxed
- Emissions are allocated on a per capita basis, using an emissions-based currency unit
- There are technology innovations and technology transfer takes place

- There is a radical redefinition of work
- More productive uses of hydrocarbons are found that do not involve burning
- International emissions peak by 2020 below 2°C, and decline to 20–30% below 1990 levels by 2050

Fragmentation Scenario

This one was an altogether more formal Scenario, focusing on the collapse of UN attempts to reach a mitigation agreement. Here are its headlines:

- There is a regime collapse as the UNFCCC fails, and the world kills the Kyoto Protocol (now after Copenhagen not so far fetched!)
- Adaptation is taken care of elsewhere in the UN system, e.g. under disaster management
- The US convinces major emitters to be part of voluntary commitments
- There is a breakthrough in technology by 2050 to address some of the challenges
- There are still voluntary, bottom-up national programmes and some regional programmes (unlinked)
- SA does take on further responsibilities, close to the 1% value scenario
- Emissions peak and decline, but much more slowly

The Montreal Case Study

In this case:

- The Climate Change regime still exists to deal with rogue states, but emissions are no longer an issue
- There is a 'Post Kyoto Protocol' with slightly deeper caps and targets
- There is an agreement for developing countries based on an agreed equity index (there are 10 years to negotiate)
- In 2030 the discussion on fossil fuels is academic as we are following a new development path
- Emissions are below or at the RBS Scenario
- SA has been a leader driving this agenda

Sustainable Development First

Here the headlines are:

- The science of Climate Change is no longer debated
- Energy demand increases worldwide, and there is increased access to energy
- Energy security becomes a major issue worldwide
- Energy prices escalate and fossil fuels continue to dominate
- We reach peak oil and probably peak gas

- Climate Change impacts worsen and population movement becomes an adaptation strategy, and adaptation actions are needed, particularly for water
- Developing country emissions overtake OECD emissions
- Big emitters have some kind of target over the next 20 years under a loose agreement with a weak compliance mechanism with voluntary targets
- Carbon has a price
- There is a realignment of the global economy, including a shift of manufacturing to developing countries
- Global stability with pockets of unrest related to resources
- African pressure on SA to act
- SA emissions grow, but we get a target as a big emitter in a SD PAMs framework
- International emissions peak at 2025, stabilise, and absolute emissions drop after this

Energy Wars

Our resident laureate, Tony Surridge presented this story, but thankfully not in iambic pentameter!

- Kyoto 2 becomes a reality
- India, China and SA negotiate to be exempted from commitments
- We pass oil peak in 2020
- There are numerous conflicts based on religion and oil (SA stays outside of major conflict)
- All agreements are overridden by conflict, and there are more local riots and conflict with time
- Climate Change impacts worsen and the focus shifts to adaptation
- Emissions increase, but not as much as GWC due to lack of development
- The economy is still fossil-fuel based
- Unless there is a stable global political environment, environmental issues will lose out
- The question is: are emissions up or down in a 'War of the worlds' Scenario?'

Nuclear Scenario

Shirley Moroka also quickly presented a utopian nuclear picture:

- 60% energy is from nuclear, 20% renewables, 20% fossil fuel
- Adaptation is not a major issue as emissions are reduced
- SA and the world's emissions are reduced
- We are meeting all our development targets, with positive economic growth
- There are no wars ...

I then proposed breaking into three groups. In each one the task was to assess one of our three trajectory Options (the ones within the framing Scenarios) against the full suite of these stories of the world in 2050.

The groups reported back after a short discussion session. What emerged was clear: *none* of the Options was *ultimately* robust against any of the future stories. The only Scenario that was robust and was plausible and consistent with all but the most apocalyptic of the future stories was the Required by Science Scenario.

The external stories had told us something fascinating: all the degrees of effort, all the actions we had modelled, would inevitably stand to collapse under the various external environments unless they fully closed the gap. We could not succeed by doing only our best, or even beyond our best. We had to do what was necessary.

'The Growth without Constraints showed us that Business-as-Usual was unsustainable, politically, economically and environmentally.' *Shaun Vorster, former Adviser to the Minister*

The Working Group that the SBT had established met for a fifth time (the Working Group had become an important part of the system, helping to crunch through the technical work much faster than would have been the case in the SBT). Much of the work now was in the final packaging of the Scenarios and Options, in preparation for the sign-off planned for SBT 6.

The Working Group was concerned that we should present a full set of true options to the politicians, not just the modelled ones. We needed to show how success could be achieved, not just how grim the story appeared to be. Laurraine Lotter suggested reviewing the language to ensure that South Africa did not appear to be a victim. It was agreed that the report currently presented a very pessimistic view.

We also looked ahead. Soon the High Level meetings would take place and the work would be presented to leaders in the four sectors: business, labour, government and civil society. The SBT process was slowly winding down.

The final draft package

Harald and I had been given the brief to present the final draft package for approval, and to package the Scenarios. With the Steering Committee, we worked on the final draft.

Once the packaging of the Scenarios was finalised, the matter of naming the three Options was a task we also took on in the Working Group. I have already referred to the names above, but the actual naming came this late in the process. A decision was made not to give the Options names that may have broader implications, or were symbolic or relied on metaphors, as is normally given to Scenarios. The names should be simple descriptors, in a form of command language. To 'Growth without Constraints' was added 'Start Now' (this path actually benefited the economy, hence the imperative), 'Scale Up', and 'Use the Market'. 'Required by Science' remained as the second framing Scenario. The formulations of 'Can Do' and 'Could Do' was abandoned, after lengthy discussion revealed (rather logically) that the dividing line between the two would be a matter of conjecture, something the SBT wished to avoid.

Economy-wide modelling

A final high-level economy-wide analysis was convened by Harald on 3 October. The participants were a stellar group: Roger Baxter of the Chamber of Mines SA; Louise du Plessis and Marna Kearney of the Treasury; James Blignaught of the University of Pretoria; Simi Siwisi from BUSA; Stephen Gelb of the Edge Institute; Michael McClintock of Sasol; the stalwart Richard Worthington from SACAN; and the reviewer, Dirk van Seventer.

This account will not analyse the work, as this is set out in Harald's book. The impacts of the various renamed Scenarios were discussed and the modelling results reviewed.

Complex economic modelling had been applied to the 'Start Now', 'Scale Up' and 'Use the Market' Scenarios. Here is a summary of the result, with the mot significant results highlighted in grey:

Economy wide impacts			
	Impact on GDP	**Employment (change in jobs)**	**Impact on poverty (change in income distribution)**
Start Now	+0,2% in 2015	2% in 2015 Jobs slightly below that of the reference case. Not large, but any job loss is of concern and would have to be off-set Lowest figure is −2,5% for semi-skilled workers in 2010	Household welfare rises 3% on average
Scale Up	+1% in 2015	Overall 1% improvement in 2015 Semi-skilled jobs peak at 3% in 2015	

	Impact on GDP	Employment (change in jobs)	Impact on poverty (change in income distribution)
Use the Market	–2% 2015 Negative effect on economy, unless off-set by other measures	Jobs increase for lower-skilled (+3% semi-skilled, 0% for unskilled in 2015) Decrease for higher-skilled workers (–2% for skilled and –4% for highly skilled)	Overall, negative welfare effects, except poorer households 0%

Some final work was commissioned and we rushed on to the next Working Group meeting, the last before the final SBT.

The last Working Group

Once again the Working Group came together, this time with some new names, for a final and very significant meeting. We presented the final draft of the LTMS and the group reviewed the work. Once again, the consensus of the SBT was confirmed. The RBS Scenario was where South Africa should aim.

I asked for final feedback in a *tour de table* format. A long list of details that needed attention emerged over the next four hours. But the work now was more about communication than substance.

The consensus was: the document was now right. However, it was immensely technical and hardly suitable for a high-level audience. Harald and I were asked to sift out a user-friendly and shorter account, as well as craft a PowerPoint presentation. With Peter Luckey's help, the full suite of documents and presentations were eventually ready.

Subject to final approval at SBT 6, our work was finished.

SBT 6: The final SBT meeting

Although a fair amount of smaller changes and tweaks were proposed, SBT 6 was effectively a ratification of the work by the SBT and its working groups, and a full acceptance of the final study report, and the reports of the four research teams. The consensus that the 'Required by Science' Scenario was the 'aspirational' future for South Africa was finally nailed down. The SBT had concluded its work.

Joanne Yawitch, the overall leader of the LTMS process, closed off the remarkable series of sessions of this unique group of people. She said that they

'The LTMS shifted the debate fundamentally. People were clearer about where things were going; but others were thinking narrowly (in their sectors), were moving more slowly and were less certain about the outcomes. Others knew that the outcomes would raise some uncomfortable facts.'
Imraan Patel, DST

were not sure whether this group would meet again and thanked the group for their valuable input to this work; she also thanked the researchers for the huge volumes of excellent work they had done. She said that she hoped that everyone would go away feeling happy with what was in the document. This had been a remarkable process of negotiation to reach consensus and this should be emphasised going forward. She said the group had managed to develop a common way forward on Climate Change domestically.

'We realised that while we are small players, we need to go the whole way, if, and so that, others did the same. Many people in the SBT were also in the international climate negotiation process. They understood these issues, and they created the critical mass.'
Imraan Patel, DST

Peter Luckey from the DEAT lead team added that the SBT members should be proud of themselves, as the information gathered in the LTMS process was of immense use. He said that there had never before been this kind of process to support a policy development process. This was a first for South Africa. The SBT volunteers had done their work. The researchers had completed their marathon. There was still some work for Harald and I, however.

We could now move forward to the High Level segment.

Planning the next phase

Work on the High Level phase had started before SBT 6, and centred on carefully refining the communication of the SBT message to the various leaders. We worked at producing three levels of communication, IPCC style: first the full set of technical reports from the SBT;[36] second a medium level report of around 20 pages setting out the policy level outputs; and finally a PowerPoint presentation, which we laboured over as the primary communication tool. This combination of the high level summary and the presentation became the primary communication tools for the process that followed. We poured a great deal of energy into preparation and communication at this point, conscious that new people would be facing up to a complex set of facts and results. It was becoming a marketing exercise!

Adieu SBT?

The significance of the work of the SBT has been referred to often. It was indeed a remarkable group that embarked on quite a voyage. But what of the results, and what of the group itself? The group may never meet again,

36 These are downloadable at http://www.erc.uct.ac.za/Research/LTMS/LTMS-intro.htm.

but in many ways it was a prototype for future processes in which evidence is collected and processed as a precursor for policy. The SBT certainly illustrated that a multi-sector, multi-stakeholder group can make progress towards a goal, and can forge a strong consensus. I think the consensus realisation that the 'Required by Science' pathway was the truly robust and feasible plan for South Africa in the long term, will be key to the future planning of our country. But the agreement remains fragile. Removing all the coal from the South African economy and then going even further remains unimaginable for most. We had agreed on the final destination, but we had no clue as to how to get there safely.

'For me the LTMS is the start. It's a start for our country. It's a country-wide discussion, and we have started the dialogue. The big question is what do you do now? What is the next step? First we need to update the numbers. Also how has the political environment changed the context, both internationally and nationally? We need to assess that. Also we need to take the themes forward practically.'
Mandy Rambharos, Eskom

Chapter 6

Response, reaction

> 'We have an opportunity over the decade ahead to shift the structure of our economy towards greater energy efficiency, and more responsible use of our natural resources and relevant resource-based knowledge and expertise. Our economic growth over the next decade and beyond cannot be built on the same principles and technologies, the same energy systems and the same transport modes, that we are familiar with today.'
>
> *Trevor Manuel, South African Finance Minister in 2008, during his budget speech*

This book is not intended to be an evaluation of the LTMS from a technical or policy perspective, but it is concerned with how the process-rich design of the LTMS may have influenced the response and reaction to the study.

The SBT, the key body to the first stage of the participation, had now completed its task. It had been a tremendous effort by a relatively small group of people, researchers, facilitators, and technical experts. The results were certainly compelling, if somewhat depressing. The study revealed that South Africa's options looked tough and our goal certainly looked distant. I was concerned that this would militate against an ambitious response once the study was taken up at policy level. The aspiration of the SBT was a result of its own journey; a journey to which future respondents would not have been exposed. I felt that our process had been exceptional in the way in which it had allowed the SBT to come to such a far-reaching conclusion. But this had been a small team, who had been close to an unfolding story.

We buried ourselves in the task of suitably packaging the study for our high level readers. We had a planned next phase of presentations to four high level groups (one from each sector), so that we could gather reactions to both the Scenarios and the aspirational agreement that had been reached by the SBT.

The four Round Tables

The Round Tables were Phase 2 of the LTMS. We wanted informality in these meetings, so we structured them as 'Round Tables', with almost café style approaches to the layout and presentation. It was planned that Harald and I would present the full report, using the communication material we had developed, with the Scenarios being my task and the economy-wide results and further elements his. We rehearsed our presentations carefully. We appointed a new facilitator, Pascal Moloi, to take full responsibility for the facilitation of the events (my task now being complete in this regard). Pascal is an affable, intelligent and widely known leader, as well as a top-drawer facilitator, and was a tremendous asset from this point forward. He could work a room and charm the socks of everyone.

Government

The government Round Table involved the Directors General of a number of government departments. This was an internal process (no facilitators or external personnel), and was aimed at securing buy-in for the study results at this senior level in the various departments. A presentation was also made at parliamentary level. The Department of Environmental Affairs and Tourism (DEAT) team meticulously canvassed government leadership.

The purpose of the Round Table was stated as:

- To provide the Directors with a briefing on the work that has been undertaken in respect of Climate Change, especially the Climate Change Long-term Mitigation Scenarios (LTMS) process;
- To request the Directors' approval for the submission of six broad policy direction themes to be addressed in a National Climate Change Response Policy to the July 2008 Cabinet meeting; and
- To request the Directors' approval for the submission of the proposed National Climate Change Response Policy development process to the July 2008 Cabinet meeting.

In the briefing, reference was made to the 'broad stakeholder consensus' within the SBT membership, describing it as:

- 'Growth Without Constraints' is an unacceptable trajectory;
- 'Current Development Paths' will not significantly change the unacceptable 'Growth Without Constraints' trajectory;
- 'Required by Science' should be our 'aspirational' goal.

DEAT officials were clearly thinking ahead.

Civil society

Twelve major NGOs, research organisations, faith-based organisations and civic organisations presented themselves at the Round Table for Civil Society. In the room were some of the most dedicated environmentalists in South Africa.

So what were the reactions?

One reaction to the study was to some degree predictable for such technical work: participants expressed the need for such a technical study to be made more user-friendly, and more accessible to the public.

Some complained that the SBT had not been inclusive enough. This later became one of the primary complaints from this sector: not enough civil society participation. Some civil society leaders have even called the LTMS a 'disastrous process', presumably from this participation perspective.

Predictably, the issue of the 'realism' of the Scenarios was also raised, especially the RBS Scenario. I pointed out that this was exhaustively debated by the SBT. Given that all the major emitters had built a broad consensus around the RBS, it was interesting to hear some civil society voices raising the 'realism' quandary.

It was also predictable that the nuclear issue would be raised, and it was. Some of the groups summarily rejected the nuclear options in the study. This formal opposition to nuclear energy in South Africa was noted. I repeated the SBT decision to model *all* energy sources. The critics will say that the LTMS presented the nuclear industry with a licence to expand.

There was a reaction that South Africa should favour a 'growth first' approach. The alleviation of poverty in the country should be a first priority, well before emissions mitigation. Harald responded that the issue of growth had been extensively debated. He said that the SBT had explicitly considered looking at a range of growth scenarios, and decided to keep within a growth rate range of 3–6% for modelling purposes. He said that the SBT had conducted a sensitivity analysis, including running a low GDP growth Scenario. He said that a broad group of economists had agreed that the structure of economy would change significantly by 2050.

In reaction to the critique on participation and breadth of representation Harald reminded the group that the LTMS process aimed to inform policy, but was not a policy-making process in itself. It was hoped that it would lead to a much more formal policy-making process with formal participation. This would include for example policy papers, the gazetting of legislation and a more detailed discussion within sectors about implementing actions.

Civil society reactions to the LTMS study still continue to be a mix of support and concern. Recent criticism is that the LTMS amounts to a short-term licence for the growth of South African emissions. Others continue to fire comments on lack of representation. Then there is the criticism of the modelling of the nuclear energy. In retrospect, the engagement from civil society is a good thing: even if some of the critique is unfair and should be directed more at the state policy flowing from the study than the study itself, the awareness and public debate must surely be healthy.

I found it interesting that initially civil society leaders did not engage meaningfully with the issues raised in the LTMS, preferring to make wider statements about growth, participation and the nuclear issue. This has changed, however.

Civil society is now, in 2010, deeply involved in energy and climate issues. The debate has deepened and is more intense, and that perhaps owes something to the LTMS. Civil society is now, in fact, leading on the thinking around low carbon planning, as I will point out in the final chapter. This year I have been working with members of civil society to develop low carbon strategic thinking, and it is clear to me that they occupy, very capably, the space in which the challenge of a carbon-free South Africa should be articulated, and that they are up there with the best internationally. The challenge for civil society is to match rhetoric with evidence, realism and practicality. Once it can do so in a comprehensive manner, it will truly lead the agenda.

'Some in civil society want to see the ideal, but it is a struggle to see how one can get from here to that ideal practically. One can have great energy objectives, but there are actually conflicting objectives. LTMS forced us all to consider the problem of how one would get from A to B. The LTMS started in the right place: how can we develop, be competitive, how do we ensure that our economy is sustainable?'
Richard Worthington, formerly Earthlife Africa

Labour

South Africa has a very powerful and influential labour movement, which is principally organised under two federations, the Congress of South African Trade Unions (Cosatu) and the National Council of Trade Unions (Nactu). A small meeting of leaders from the federations was convened in the third Round Table. The management team presented the material from the SBT. The debate moved swiftly into the negotiations territory: labour, rather interestingly, was most interested in how the LTMS would serve to inform and shape our negotiations in the UNFCCC process. I would have expected the leaders to pour over the job-creation results, and the impact on the coal industry! Perhaps they were way ahead of me on those issues.

Interestingly the comments also pointed to the long-term nature of the LTMS, and the problem of implementation over such a period. Then of course the issue of declining industries (and in the background, employment levels) was also raised. The very real interests of labour in growth, job creation and poverty alleviation, were articulated.

There was clearly a tension between 'green issues' and the need for development. By 2010 this has changed significantly, with the green economy now central in the growth debate.

It is interesting to me that in neither Round Table was there a deep engagement with the issues (at least not in that session), and I am inclined to think that we presented things far too rapidly, with no process to allow for engagement with the results. The seemingly tepid responses and limited engagement was, for me, a sign that the packaging of the material fell short of what was required by these two sectors. It became clear that we need to have raised capacity to engage with the issues in a far less technical way in order to take the energy/Climate Change issue to a totally new level of discussion in society. This is work for upcoming years. This is the effort of full social accountability.

Business

The reaction in business was a different matter altogether. For business, emissions are (in our energy economy) an unavoidable part of doing business. The LTMS addressed this head-on.

In the case of business, we wanted a good number of the CEOs from the top 50 companies to attend, and we also wanted the CEOs of Eskom and Sasol, as the two largest emitters in South Africa, to be there. Finding a suitable date, and securing the attendance of CEOs (in any country) is like herding cats. These people have hectic diaries, and characteristically don't come to meetings unless well prepared, or suitably drawn by the importance of the agenda. They lead the largest companies and emitters in South Africa, however, so we needed them to come. We decided to ask our Minister to invite them in person.

With the help of the leading business organisations, we came up with a list of some 40 CEOs and sector leaders, and then searched through their organisations for the people who, on a technical level, advise them. We then called these advisers to a preparatory meeting, and gave them a full briefing on the LTMS results. We expressed the hope that they would engage with their leadership and properly pre-brief their CEOs; we also implored them to secure the attendance of their leaders.

The Minister of Environmental Affairs and Tourism then called the meeting. Thankfully the preparation paid off and a good turnout of CEOs was recorded.

A special moment

Harald and I were ready to address this group of top business leaders on the night, albeit with some dry-mouth nerves. Pascal was as relaxed as ever, and knew most of the people personally. It turned out to be an easy night for him, as the Minister took over facilitation of all the dialogue in his characteristically affable way.

I looked at the group, recognising the faces of some of the most distinguished, intelligent and renowned business leaders in the country. Harald and I knew that our presentation of the LTMS had to be spot on. Unlike the other two groups, where we could anticipate a degree of support for the study, with this group a total rejection of the work was equally possible.

Minister van Schalkwyk engaged with the group, assuring everyone that shortly he would take the LTMS study to Cabinet. Before doing so he wanted to hear their voice, as he wanted to include their response. He emphasised how important the report was, and that they had his commitment to obtain an appropriate Cabinet mandate on this issue.

He then asked for comments.

The first comment received was for me perhaps one of the most significant inputs by a stakeholder in the entire LTMS process. The CEO concerned gave a five-minute input, which was as impassioned as it was eloquent. South Africa, she said, had no option but to *plan* its economic development to align with the 'Required by Science' Scenario. Picking up on the same theme which the SBT embraced in the final SBT sessions, she pointed out that any other course of action was untenable for the country: hence it was essential to plan with this emissions trajectory in mind. She was moving ahead of the LTMS into the next phase: the planning of a low carbon economy. She added that the challenge was not only to align the economic development of the country in this way, but also to align the efforts of government, of government departments, of business, the utility and emitters in the economy, to ensure that their collective efforts, and the collective policy and market signals, would track the emissions along this trajectory. Aware of the immense challenge inherent in this Scenario, she nevertheless exhorted government to declare itself linked to this Scenario for South Africa. The CEO asked the entire group of leaders to support her in this call, and one by one they added their endorsement.

'The Stern Report and the McKinzie Report seemed to be the turning points in the UK and in my mind in the South African context that turning point was the LTMS. For the first time the Climate Change challenge became quite clear to us, how daunting it is and more specifically what the options were for South Africa to pursue in an integrated way. The LTMS showed us the complexity of the challenge, the scale of the action needed, and the integrated approach that needs to be followed.

The LTMS was one of the things that ensured that CEOs really started understanding the Climate Change challenge in South Africa.
The most important element was awareness and our awareness that not any one technology could solve our problem and in fact to some extent we needed miraculous intervention.'
Herman v d Walt, Fred Goede, Sasol

I was truly astonished. The Minister had his mandate, at least informally. The extraordinary consensus, which to a degree amounted to a step into the unknown, was building around the LTMS study.

Business reported that it needed to go to its constituencies to get a formal mandate. We would have to organise a second event to receive the formal response.

The presentation by Business Unity South Africa (BUSA) at a second round of the business leaders' engagement was significant in that it underlined the informal position taken at the first meeting. Amongst other detailed recommendations, it asked that:

- All proposed options be elaborated as part of strategic national Climate Change response;
- Further research was required in consultation with stakeholders;
- A coherent intergovernmental approach should be followed; and
- South Africa needed an action plan rather than more policy.

It recommended a high-level continuation of dialogue between government and business. It recognised the extent and sheer challenge of the task – but maintained that this was the 'aspirational goal' for South Africa.

'It was interesting to see the CEOs talking like activists. I don't think anyone knew what all this would mean, but despite not being fully briefed on the impact of this ambition, they took the ambitious route. To the Minister's credit, he built a high-level consensus.'
Imraan Patel, DST

'We have great thinkers, but we get stuck with implementation. But LTMS showed that it can be done. The discussions were not easy. We are often irritated with each other. But we got it done.'
Mandy Rambharos, Eskom

The Cabinet response

The last of the Round Tables had taken place in April 2008, marking the close of the LTMS study. Harald and I were 'sent home', having done our jobs, along with the facilitators, secretaries, stakeholders, researchers, department officials, caterers and support crew. We now awaited the official response.

The Cabinet response to the LTMS was released in July 2008, exceptionally rapidly. It was comprehensive, and attested to a tremendous preparation effort by the Department. One does not know what happens in these meetings (and no-one tells) but the response hints at the debate that must have taken place.

The full text of the media release is worth reading,[37] as well as the presentation that followed.[38] The response came in the form of a set of policy 'recommendations'. Six themes were outlined:

Theme 1: Greenhouse gas emission reductions and limits;
Theme 2: Build on, strengthen and/or scale up current initiatives;
Theme 3: Implementing the 'Business Unusual' call for action;
Theme 4: Preparing for the future;
Theme 5: Vulnerability and adaptation; and
Theme 6: Alignment, coordination and cooperation.

The highlights for me are in Themes 1 and 3.

On the issue of the future of business in South Africa, the Cabinet echoed Richard Worthington's statement: business as usual will put us out of business, and it called for business 'unusual'. The 2010 Green Economy summit is a follow up to this agenda. Although the green economy is still a new, and perhaps only partially understood concept, the LTMS pushed it onto the stage, and read its first lines.

On the issue of emissions, the Cabinet recommended that South Africa present the world with a 'Peak, Plateau and Decline' Emissions trajectory. This was presented in the illustration on the next page (the light grey line).

This is the central and extraordinary response to the LTMS study. It requires some understanding. Its significance to South Africa's future seems set.

Firstly, this is not a technical response, but one that looks at the headline significance for South Africa. It is devoid of specific detail, of figures and limits, but it does hint at particular thinking and planning. Hence it should not be greeted with disappointment as a vague and general response, as that would be to fail to understand its significance.

---

[37] See Annexure 10.

[38] See Annexure 11.

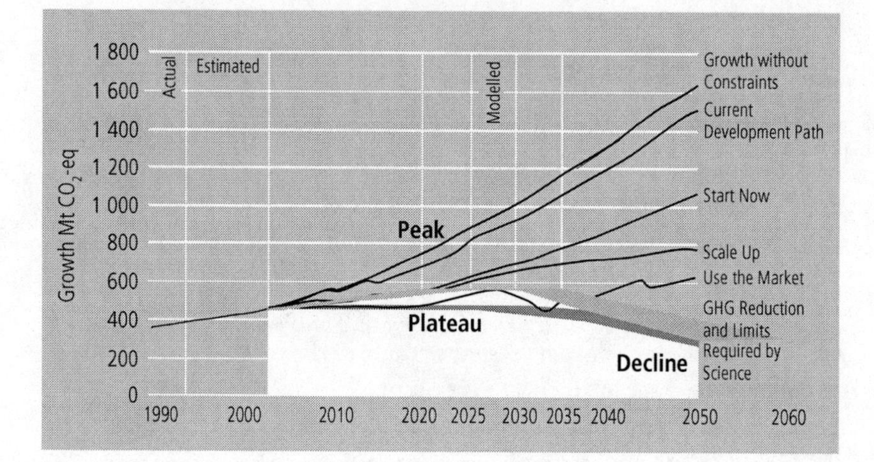

Secondly, the Cabinet's recommendation is incredibly ambitious for South Africa. As an emissions reduction plan, it was perhaps one of the most ambitious in the world at the time, even more so, relatively speaking, than some of the industrial countries. To commit South Africa, even broadly, along this emissions path is to retool or re-engineer our economy and our society in a remarkably comprehensive manner. The last 30 years saw our economy developing more or less as resources were discovered and exploited. Now a new driver would be on the scene, and it would require a far more active approach to our economic development.

On analysis the early phase of the trajectory carves out more 'carbon space' than set in the RBS Scenario, ostensibly to create some slack for a coal-based solution to the 2008 South African energy crisis and to medium-term demand. The construction of the two large coal-fired power stations – Medupi and Kusile – account for the higher peak. This has become controversial, but it must also be remembered that the plateau means the end of the development of any further coal-fired power in South Africa. After that it means the steady, and then rapid decommissioning of this energy source.

The rise to the peak is squarely based on the 'Start Now' Scenario, but the departure from the Scenario in 2020/25 is just around the corner, and hence urgent and real preparation will be key to making the turn when it arrives. It will have to be the decisive decoupling moment.

There is also still considerable risk in the proposal. We know from the LTMS that there is no way, with today's known technologies, that we know how to reach the goal in the RBS, which has the same endpoint, roughly, of the decline trajectory. So is the decline a leap of faith? Clearly in the minds of the Ministers there was the hope that by 2030 and beyond that there will be the required technologies, that new resources will have been found and exploited, that so-called 'soft' measures will be instituted successfully, and that the transition to the low-carbon economy will be well on its way.

Some cynical voices say that there is no intention, really, to achieve this at all, and that the ANC is a 'short-term' government which created space for the two new coal-fired power stations to be built in the peak era, and that there is no real intention to achieve the plateau and decline. This will be tested both at home as an effort of long-term political will, and abroad in the negotiations. So far there is evidence of total commitment, despite a change in the Presidency and the Cabinet.

The next question is: without figures and limits to the peak, is the trajectory so elastic as to confound any effort to measure it, or to bring players into compliance with it? The answer, I believe, is that the country is not ready to accept a quantitative emissions trajectory for the economy. We simply don't have the measurement tools and the ability at this stage to track the economy along such a trajectory with any degree of success. For that you need, for a start, a sophisticated inventory system, confidence in the compliance measures, and a carbon plan combining market and implementation interventions. We don't have that yet and putting figures to the curve at this point is premature. At the same time some will exploit the slack and we can expect a lot of arguments once the peak in absolute emissions for South Africa does enter the debate.

The international stage

The Peak/Plateau/Decline approach is the central reaction, or result of the LTMS study, and has become significant internationally. It summarises the government response to the LTMS. It is a compromise between the rigour of 'Required by Science' Scenario and the pressures of the energy crunch in South Africa. The recommendation was ready for presentation to the international community at the end of 2008.

The LTMS had already influenced the South African position at the Bali UNFCCC Conference of the Parties at the end of 2007 – it enabled South Africa to 'take the high-road' internationally and demand that others do so as well. But the LTMS was truly presented in full for the first time at the Poznan COP in 2009, where it was met with acclaim. There was tremendous interest and appreciation for the Peak/Plateau/Decline decision, and much interest in the stakeholder-driven process.

Reactions were widespread at the time, but I focus here on the reactions of some of the SBT members.

'The impact of the LTMS was amongst other things that it gave South Africa a lot of weight in the international negotiations – weight that it did not necessarily have before. It is also a very useful document to have in hand to support the evolution of our policy approach to the climate challenge. Without it we would have been a lot worse off. It was a very timeous document.'
Guy Midgley, director SA National Botanical Society

'The LTMS was presented (at the COP in Poznan) and most people were immensely impressed with the process, asking: "How did you do this?" I guess this is how we do things as South Africans.'
Mandy Rambharos, Eskom

'The Peak/Plateau/Decline approach was a direct result of the LTMS study, and is now an international commitment. That was very significant.'
Richard Worthington, formerly Earthlife Africa

'The Peak/Plateau/Decline represented a good tactical compromise to the strategic objective set in the Required by Science Scenario.'
Bob Scholes, CSIR

'LTMS has been of fundamental importance to the way we interpret the contribution [to emissions] of developing countries in the future – this is the international aspect. It is also one of the first instances that I have seen that a developing country has gone to such lengths to transparently develop what it could and should do about GHG emissions. I have been in meetings where the question is "How do we localise the LTMS? How do we take the LTMS to the next level of action? And turn it into an implementable action plan".'
Kevin Nassiep, CEF

'The LTMS was legitimated both by government and by the participation of key players.'
Bob Scholes, CSIR

'There were two key interventions that the LTMS achieved: running the process diffused some building tensions between Eskom and other components of South African society; and secondly it radically informed South Africa's international negotiation position – not only South Africa's own position but the space it took within developing countries.

These two outcomes are more than adequate.'
Bob Scholes, CSIR

'International meetings always refer to the LTMS concept of Peak, Plateau and Decline with tremendous interest and respect: it is a pity that we have not recorded just how much this new concept impacted upon international thinking.'
Kevin Nassiep, CEF

'Business learnt that is possible to sit at the table with civil society and learn something. The LTMS was a unique exercise in participation.'
Richard Worthington, formerly Earthlife Africa

The World Bank review and other responses

The LTMS was peer reviewed in July and August 2008 by a team from the World Bank. The international review team, comprising Xiaodong Wang (Task Team Leader), Emilio La Rovere (Energy and Climate Change Expert), Ming Yang, (Energy and Climate Change Expert), and Catherine Fedorsky (Southern African Energy Trade Expert), visited South Africa from 30 June to 4 July 2008, to hold a workshop and provide peer review comments to the Long-Term Mitigation Scenarios.

Here are some of the excerpts from the review team's comments:

> Overall, the review team believes that the LTMS is the first of its kind in developing countries with South Africa a leader in this area. The team found that the combination of research-based Scenarios with stakeholder consultation process was a pioneering effort to provide high-quality information for decision making on Climate Change response strategies in South Africa. The methodologies used in the research were consistent with international best practice and the results are robust. Notwithstanding the potential improvements that could be made in future work, the results clearly provide the basis of best available scientific information for decision-makers. The work lays a robust basis for development of domestic sectoral implementation efforts. The energy modelling underpinning most of the analysis was found to be robust. The dynamic economic modelling should be completed as a matter of priority. Already, the existing analysis enables proactive planning for a transition to a low-carbon economy. The innovation shown in the LTMS Scenarios would be worth sharing with other developing countries.
>
> The LTMS process deserves continued support, unfolding into implementation. The team was quite impressed with the stakeholder consensus on the 'Required by Science' Scenario, as well as the recent media coverage from the DEAT Minister on LTMS. However, the review team would like to highlight the huge challenges ahead for implementation.
>
> Finally, the team suggests further studies to include: (1) estimating the cost of inaction (e.g. a national 'Stern Report') to illustrate that action now on mitigation and adaptation is cheaper than delay, (2) coupling dynamic economic modelling with energy

> modelling, and (3) sharing LTMS experience with other key developing countries such as those in the Southern African region, Brazil, China and India.

On the process element, the team had the following to say:

> The team is most impressed by the rigorous stakeholder consultation and consensus building process leading to LTMS and believes this is a major achievement of the exercise. It started from an endogenous initiative in South Africa and unfolded into the involvement of key stakeholders in the Scenario Building Team (SBT). Different governmental agencies, representatives from the business world (see CEO list of participants) and NGOs were members of SBT, besides technical experts. The many rounds of meetings and discussions held in the process are essential to build consensus among stakeholders, as Climate Change mitigation involves many sectors and stakeholders across the country and has significant implications to the economic growth and development prospects. Particularly impressive is the consensus reached among the senior level decision makers for the Required by Science (RBS) Scenario. The key strength of the consultation process was the consensus that stakeholders achieved at the input stage built the platform for the same consensus at the results stage; which in turn strengthened the credibility of the result at the political level. This pioneer effort was so relevant that it deserves continued support and development.

The concluding remarks made were:

> LTMS provided valuable inputs for decision-making on Climate Change mitigation policies in South Africa, based upon sound scientific analysis and the involvement of key stakeholders.
>
> It is strongly suggested to provide additional support to this initiative in order to pursue LTMS effort and further extend its scope.

And finally the suggestion was made to:

> [Share the] LTMS experience with other key developing countries such as those in the Southern African region, Brazil, China and India. The South African experience in facilitating this process can be very valuable to these countries, and technical exchanges

on modelling and Scenario design may be mutually advantageous (benefit from the Low Carbon studies supported by the World Bank in these countries).

On balance

Lavish praise, or damning criticism? I have heard the words 'unmitigated disaster' used for the LTMS. No process is perfect, but the general consensus view is that the LTMS got it right.

> 'The process contributed immensely to building awareness in business. There are now few high emitting businesses that are not doing something about greenhouse gas mitigation.'
> *Laurraine Lotter, CAIA*

Some criticisms

Some helpful critique has also come to light over time. In talking to stakeholders, it seems that criticisms can be grouped under the following headings:

1. The time things took: in this regard I have commented elsewhere and this is soft criticism, as most were aware to what extent this was a primary process for South Africa. Imraan Patel also noted that the LTMS needed 'time to mature' and that the time it took was an advantage.
2. The modelling of the carbon tax: this is a much more trenchant criticism coming mostly from business and starting within the SBT. The critique at the time was that modelling a carbon tax as part of a group of mitigation actions was an erroneous approach as a carbon tax was not a mitigation 'action', but rather a market instrument (among many) that could incentivise mitigation actions (which were already modelled). This is of course correct thinking and so the 'Use the Market' Scenario does stand out as an out of place picture to some. At the same time it is just simply what it is – a picture of the effects of 'what if we used a tax?' I think the agreement in the end (despite some bitter arguments in the SBT) to model the tax was perhaps gained out of curiosity: a general sense of 'we don't agree that it should be put there, but we'd like to see what it will do', and a somewhat surprising result that it gave.

> 'The LTMS got people comfortable with the concept of a Low Carbon Economy. They understood that it is an inevitability. It ceased to be an "NGO dream"!'
> *Richard Worthington, formerly Earthlife Africa*

> 'The Carbon Tax should not have been modelled, as it is not a mitigation action. The reason it was modelled, as to see just how big a difference it would make, and it did in the modelling. But in real life it may just be a revenue-making exercise. We don't ring fence taxes. Again this is why we need the follow-up dialogue – we would have to recycle such a tax.'
> *Mandy Rambharos, Eskom*

3. There is a further criticism, though, which is that the modelling of the economy-wide impacts of the tax, was insufficient to give us a real understanding of what the effect of the tax would be, yet it (the LTMS) has been read as a mandate for the implementation of such a tax. This is partially a fearful reaction from industry, especially those who may face heavy carbon taxes, but it is also a real concern for the general economic impacts the tax might have.

   There are a number of issues here: in the first place the Scenario modelled not only a tax, but a packages of incentives as well, and these have been 'lost in the debate'. Secondly, it must be admitted that the LTMS is not a clear and final study on the tax issue. Thirdly, South African Revenue Services have not agreed to any ring-fencing of tax revenues: the LTMS Scenario is based on recycling the revenue. Thirdly, and perhaps most profoundly, the whole point of a carbon tax is *not to have to pay it* – only then is it actually working, as this means that it has pushed the uptake of alternative, tax-avoiding, energy services to producers and consumers. The signal of the tax is, in some ways, more important than the actual payment of it, at least at first. The tax must come eventually, although many think that putting a price on carbon is merely another exercise in neo-liberal economics, when the real low carbon interventions should be executive action across the system.

> 'What is unfortunate is there has been a temptation to take a piecemeal approach to carbon, and the carbon tax has fallen into this bracket. The document is the base for a Climate Change response strategy and policy; hopefully we will see the concerns that we have raised in the process taken aboard in the development of this policy.
>
> I don't think you can adopt a carbon tax without fully understanding the impacts, and I don't think LTMS fully developed a socio-economic understanding of the tax.'
> *Laurraine Lotter, CAIA*

4. The 'support' for the nuclear approach: the LTMS modelled nuclear energy and renewable energy, based on a stakeholder agreement that even-handed information on both energy sources was better than favouring one over the other. For the anti-nuclear lobby, any 'acceptance' of nuclear energy and most references to costs of nuclear energy will be subjected to heavy criticism, for all the obvious reasons. The fact that announcements by the government about 'fleets of nuclear power stations' have followed the LTMS, will no doubt further discredit the LTMS as a whole.
5. The exclusion of representatives of civil society: I have already mentioned this issue, and attempted to justify the method we followed in

selecting members of the SBT. The High Level segment was more open, but it may have been that significant players had already withdrawn from the process at this stage. The criticism that the High Level segment favoured the business sector must also be noted. But more strongly, if one equates the LTMS with a policy-making process, then these criticisms are correct. But as it is not, the limited approach to consultation and involvement is (at least in my view) justified. The LTMS was a step towards developing social accountability around the development of a low carbon South Africa: merely the first step.

> 'I still think we didn't get enough people for the high level segment – I wonder if the results would have been different had we involved larger numbers of CEOs. It may also have been interesting to have brought the sectors together at the high level segment. That combination of sectors would have tested the depth of the consensus.'
> *Kilebogile Maroka, DEA*

6. The way the message came out turns out to be something of an unusual and unintended consequence. I will try to sum up this point as follows: when the LTMS report was released, and shortly followed by the Cabinet decision, there was a jolt of climate awareness in the South African system, and the LTMS rapidly became something of a household word in leadership circles. Business had, in fact, insisted on a follow up process to the LTMS: this was to be a high level dialogue in which the framework for the action plan and the supporting policy package would be created. But this has not happened yet, and South Africa has had to endure a diverting (and bruising) political change of guard since the LTMS. This has had all sorts of odd consequences: the LTMS has been seen by some as the product of the previous, largely discredited dispensation, and not judged on its merits. But larger in the critique is the fact that the LTMS subtly turned from study into prescription. This is perhaps because of the vacuum of action that followed it, but also to its own internal persuasiveness as a study. A victim of its own success, perhaps. One has to constantly recall that the LTMS was a study in possible future mitigation stories – a set of Scenarios. The policy development and executive decisions that follow are made of different stuff.

Finally, Copenhagen

The LTMS really paid for itself when the government made its emissions pledge at the Copenhagen COP in 2009. The pre-COP press release from the South African President's Office reads as follows:

For South Africa, the major contributor to our emissions of Green House Gasses is our energy sector. However, the issue for developing countries like ours is not merely about addressing our Green House Gas emissions but also about energy security and energy access as well.

The greatest challenge we face is how to ensure both energy security and access as a developmental imperative and at the same time laying the foundation for moving towards a path of low carbon growth. In the short to medium term we have an immediate energy supply challenge which alternative energy supply options cannot meet at affordable cost and at the scale needed, therefore, we are aggressively pursuing carbon efficient coal technology, in the medium term.

The science is very clear – there is no 'silver bullet' – Climate Change is a huge global challenge which will take a combination of the full range of available interventions, technologies, policies and behaviour changes to resolve the climate problem. It will also demand massive investment in new low carbon technologies. Economies across the world have to put long-term plans in place to transition towards a low carbon growth path. In this context, we have modelled South Africa's mitigation potential and potential low carbon solutions in the Long Term Mitigation Scenario (LTMS) study. This work is being used to inform the policy choices that will allow us to aggressively address Climate Change in a way that unleashes the job creation and developmental opportunities of a 21st century 'Green Economy'.

As such, South Africa, being a responsible global citizen and in line with its obligations under Article 4.1 of the United Nations Framework Convention on Climate Change acknowledges its responsibility to undertake national action that will contribute to the global effort to reduce greenhouse gas emissions. In accordance with this, South Africa will undertake mitigation actions, which will result in a deviation below the current emissions baseline of around 34% by 2020 and by around 42% by 2025. This level of effort enables South Africa's emissions to peak between 2020 and 2025, plateau for approximately a decade and decline in absolute terms thereafter.

This undertaking is conditional on firstly, a fair, ambitious and effective agreement in the international Climate Change negotiations under the Climate Change Convention and its

> Kyoto Protocol and secondly, the provision of support, from the international community, and in particular finance, technology and support for capacity building from developed countries, in line with their commitments under both the Framework Convention on Climate Change and the Bali Action Plan.
>
> The potential for multilateral finance to unlock ambitious mitigation actions is already evident in recent events. For example, South Africa's successful application to the Clean Technology Investment Fund has successfully mobilised $ 500 million, leveraged to over $ 1.6 billion from other multi-lateral sources in order to support the establishment of a i) 100 MW utility scale wind power generation; ii) 100 MW Concentrated Solar Power Plant iii) conversion from electric water heating to solar water heaters for 1 million households; and iv) scaling up of energy efficiency projects as leverage for commercial and industrial sectors. Clearly the scale of support enables a concomitant level of action.
>
> In this regard, South Africa emphasises that an ambitious and long-term financing package for both adaptation and mitigation is a central element of the Copenhagen negotiations and one that will have significant impact on the extent to which developing countries can take mitigation action.

That response is well captured in the statement of the new Minister in the environment portfolio:

> We have at the highest level of government agreed that our emissions should peak by 2025, plateau for a decade and then decline from 2035. In order to achieve this, we intend to implement sustainable development policies and measures, including shifting to energy efficient technologies, and using our abundant solar and wind potential to roll out renewable energies on a large scale. We have demonstrated, through our work on Long-term Mitigation Scenarios and through our assessment of our nation's vulnerability to the impacts of Climate Change, our willingness and readiness to mitigate our emissions and to take action to adapt to the impacts of Climate Change. However, without financial and technology support, it will not be possible to do more than what we are already doing.[39]

[39] Speech by Ms Buyelwa Sonjica, MP, Minister of Water and Environmental Affairs at the National Information and Consultation session at Collosseum, Pretoria, 10 November 2009.

Chapter 7

From Scenarios to plans

History will tell us if the UN process has succeeded in meeting the climate challenge, and ultimately the degree with which the LTMS influences action at home in South Africa. There is much debate internationally on whether we will ever reach the 'big global agreement', or whether we should focus on increasing ambition domestically and in cluster agreements. The question at this stage for me is whether the players in South Africa will stick to their resolve, and will remember their 'remarkable consensus', their robust agreement, their ambition and their vision for our future. Put another way: the question is whether the process followed in the LTMS will succeed in creating permanent resolve, impervious to the short-term demands of the market, the shifts of politics, and the demands of an energy-hungry consumer. Will it have made a difference? Will its lever for action continue to hold?

> 'However when it comes to the level of ambition this requires, the costs, the actual steps – this is where we are not yet on the same page. Our expectations of how this would play out were quite different. And we don't yet have a sense of how the South African public will react to plans to tackle this mitigation challenge – we live in a spoilt society where people expect low cost electricity and aren't efficient.'
> *Mandy Rambharos, Eskom*

There are some signs that business is still kicking up against the drive to a low-carbon economy for South Africa, but most signs are that we are sticking to the curve. In some ways resistance is understandable: industry needs constant available energy. There are also signs that consumers still want cheap, constant and abundant coal-driven electricity.

Coal dominates the agenda. Some see the ongoing reliance on big coal as amounting to a bad decision for the country for reasons other than emissions, reasons that show up gaps in the 'emissions-based' approach:

> ... the Cabinet has bought the Minister of Energy's mistaken view that South Africa has 'carbon space' to invest in coal-fired power stations (this, despite the fact that South Africans emit 9.8 tons of CO_2 per capita per annum, the

> same as the UK). From a carbon perspective, the numbers do not support the conclusion that we are a developing country. In reality, however, this decision should not be about carbon or the environment at all, but should be a purely economic decision, i.e. what will best incentivise innovation, job creation, investments and global competitiveness? Investing in mature capital-intensive technologies creates less jobs than investments in new decentralised technologies that are not captured by monopolistic value chains. South Africa needs to harness its entrepreneurial energies to drive returns on innovation and not just rely on the same old SA story – cheap resources and labour exploitation. Countries with the most diversified economies in the world are also the countries with the most expensive energy. Why? Because as energy prices go up, innovation for diversification follows. Eskom has never understood this simple fact, hence the crazy idea of providing its biggest customers with the cheapest electricity.[40]

One thing is for certain: since the LTMS there has been a great deal of public debate, and increasing calls for more and more transparency in an arena once almost totally closed to society: energy. The Climate Change initiative acted like a wedge in the crack. But the debate is still rather crude in many ways, and we are far from having a clear plan.

This of course takes me back to the origin of this book: how does one best involve society in the 'mitigation decision', which has long been too technical for policy makers, and too political for technocrats; too confounding for the consumer, and too threatening to the investor?

The LTMS and the policy process

The LTMS did what it was primarily supposed to do: it helped the government to formulate a response in the international negotiations. But it can be argued that it also did something much further reaching: it launched the debate in society. By 2010, the LTMS as a study has transformed into the LTMS as our Climate Change Plan and is literally under discussion everywhere in South Africa: boardrooms, newspapers and business schools.

> 'One of the problems with LTMS is that people read the results without understanding what went on behind them and start turning Scenarios into fact.'
> *Herman v d Walt, Fred Goede, Sasol*

[40] Mark Swilling, *Cape Times*, 7 April 2010.

But this needs to be placed in perspective.

To some extent this prominence was also a problem for the LTMS – its identity was steadily being conflated with policy and the direction government was taking on Climate Change. It has become a synonym for our plan, which means that for some it is the villain of the peace, for others the panacea.

The LTMS was really only the start of a dialogue, primarily between technical leaders and decision-makers. The new dialogue is about planning, and this is where the LTMS reveals its shortcomings.

'We almost need another process now, to talk about the details of the plan forward. We need another SBT. It won't be easy! But the same structure would be so useful to unpack these issues. To try to achieve a national discourse again. Without this, the LTMS starts to look like a strategy, rather than a study.'
Mandy Rambharos, Eskom

Planning for low carbon

In recent months a number of colleagues[41] and I have looked again at the LTMS, and asked ourselves: is the LTMS a useful tool in planning for a low carbon economy? The answer seems to be both a yes and a no.

The LTMS, and many other exercises in emissions modeling worldwide, focused on a method that essentially relied heavily on a base case – a future constructed through modelling. Mitigation interventions are then compared against the base case. However, in recent months these emissions base lines have come under some fire when a number of developing countries, including South Africa, used them as a marker for the pledges that they offered in Copenhagen. The concern is that the base lines are intentionally inflated. Climate Tracker commented on South Africa's target (see the final sentence):

'It comes up everywhere that the LTMS is South Africa's strategy on Climate Change; it must be remembered that the LTMS is not a strategy. It was a study.'
Kilebogile Maroka, DEA

'Because it wasn't a policy document it became possible to take the high road – but the reality is that when you now try to convert it to policy you are not going to get the high road, certainly not in the immediate future.'
Laurraine Lotter, CAIA

> South Africa is rated Inadequate. It provided a quantitative target based on a detailed, comprehensive Scenario analysis, but has made the target conditional to a strong Copenhagen agreement and financing from developed countries. It is unclear which share of the target is unconditional. If the conditional target becomes unilateral it will be in the Medium range. The reference

[41] With support from the World Wildlife Fund, and with acknowledgement to the team.

> Scenario that is related to the (Copenhagen) pledge is considered relatively high.[42]

The validity or the base-line approach, and for that matter its ability to be accurate in tracking the country's emissions against a planned trajectory, depends on the accuracy of the base line. However, as we know, the base line is a modelled version of the future, and relies heavily on assumptions which, in most cases, will be trumped by reality. Even if we constantly review our predictions, we don't really have a full-proof planning tool.

A way of avoiding this criticism and the planning deficiencies of the baseline driven approach is to adopt another possible approach: using fixed carbon amounts, or carbon budgets. In the former method your pledge is made against a fiction, whereas in the latter it is made as a permanently fixed amount over time.

Internationally we need to keep concentration of CO_2e below 450 parts per million for a 50:50 chance of keeping global mean temperature below a 2°C increase. The emissions pathway to ensure that the 2°C global mean temperature limit is sustained (subject to the uncertainty referred to) has been calculated, and in effect provides us with a global carbon budget. Research by groups such as Ecofys[43] and others tell us that on this century the global carbon budget is around 1 800 gigatonnes CO_2e (assuming land use change and forestry becomes net absorber of emissions by 2100). Our emissions between 1990 and 2008 have already used around 40% of this total budget, and hence we now have a restricted budget before 2050 (the horizon year in the LTMS) of around 950 gigatonnes CO_2e.

'There was a definite context timeline behind the LTMS. Its timing was spot on. The IPCC process was just being recognised as credible, that context had matured.

Policy is never finalised. I think there are still major battles to come. The LTMS reduced the differences. But we still have to question how to get onto the Peak, Plateau and Decline trajectory. So: where to from here? What processes would work best? How do we maintain the LTMS commitment, prevent it from being dissipated? How do we put together the actual plan?'
Imraan Patel, DST

'The LTMS helped leaders start repositioning now in preparation for a carbon-constrained world, a world where there is a price on carbon.

I think that the LTMS contributed to an understanding of the global burden-sharing implications for South Africa, as the world tries to stabilise the climate at 2 degrees.

We had very progressive industry reaction to the LTMS: a desire to be leaders in the low carbon economy.'
Shaun Vorster, former Adviser to the Minister

---

42 http://www.climateactiontracker.org/country.php?id=1868.

43 http://www.climateactiontracker.org.

The next step is the division of this budget on a national basis. Graphically, this can be illustrated as follows:

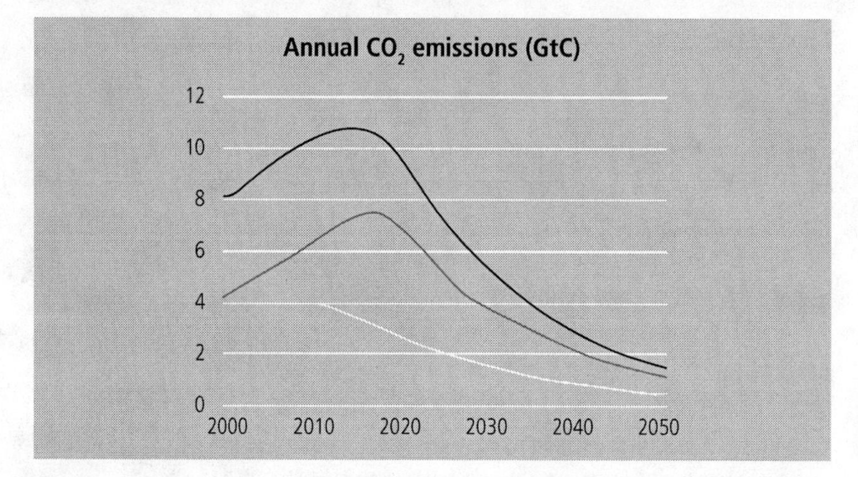

Here the black line is the global trajectory, with a division between the developed (white) and developing (grey) worlds. Together this sets us on course to be carbon-neutral before the end of the century. The space below the line is the carbon budget.

The budget for South Africa is a piece of this larger budget. Calculating this budget is contentious, to say the least. Here is one possible way: South Africa's share of the world total of absolute emissions is 1.6%, our population is 0.73% and our economy is 0.71%, so let's say that our budget (given some more development space) settles on 1%. This would mean that we have a budget of 9.5 gigatonnes for the period 2002 to 2050. However, when we look at the South African pledge, we see this picture:

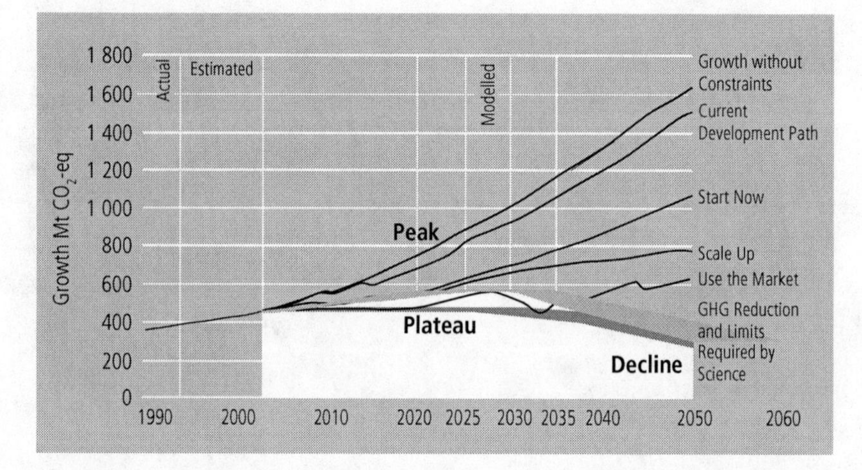

Here the carbon budget (the white area) can more or less be quantified at between 15 and 18 gigatonnes. This is double the 1% budget. Doubling the budget means that another country or countries needs 1% less. This means that there is a considerable developing country discount (an assumption that developed countries will take up the 1%) in our RBS Scenario. Of course applying such a discount to all developing countries blows the global carbon budget. We are not alone in wanting more carbon space, and others talk in much bigger numbers.

This is essentially why the world cannot come to the 'big agreement'. The figures don't add up. We cannot balance a budget with 2°C that is easily shared between nations.

Either way, what this does mean is that South Africa will *have* to achieve its set goal, and that urgent planning to achieve this is necessary. But how would such planning be structured, given the fact that through the LTMS we know just how difficult this will be to achieve? Therefore, how can we structure the planning process, learning from our success in the LTMS, to achieve the following goals:

1. An understanding of how the budget could be divided between sectors in the opening years, based firstly on evidence, but also on factors such as cost, ability, pace and impacts on jobs and poverty. The construction of the evidence base for this understanding is a development of the work done in the LTMS.
2. An involvement by stakeholders, particularly in the sectors where emissions are significant, on both a sectoral and inter-sectoral basis, starting with an acceptance that we have to find a division that works within the budget.
3. An understanding of how one could drive the result, using combinations of instruments and approaches, and constantly monitor the results, enabling ongoing adjustments.
4. An understanding of what the policy instrument that will drive this will look like.
5. As the peak approaches, and our efforts will have to increase, we will need to understand which major infrastructural activities will be required, especially in the electricity and transport sectors, where our emissions are most significant. The lead times and impacts of these build and development programmes will have to be factored in.
6. Early preparation of the 'fairy godmother' technologies and shifts, both social and economic, need to happen. We need to understand

the evidence of consumption-based emissions modelling, as against production-based emissions data, and we need to get these two approaches to 'meet in the middle'.

7. From a stakeholder perspective, the really hard task concerns the role and future of the large emissions heavy industries. For example, the total dismantling of the coal industry and its related coal driven power stations is almost a given, unless we can solve the capture and storage problems within a set period. How will stakeholders react to this reality, and plan accordingly? What impact will this have on jobs and livelihoods? On the stock exchange? On exports? How will we best design the participation, so that stakeholders can engage and design their futures without fear or withdrawal?

The 'low-carbon South Africa' project would have to include these eight priorities:

1. Doing things more efficiently at all levels;
2. Providing energy services that are eventually virtually carbon-free;
3. Facilitating human and product mobility, in a way that is also virtually carbon-free;
4. Totally rethinking housing and buildings, and the way we plan space;
5. Focusing on food and land issues;
6. Restructuring the economy;
7. Restructuring society; and
8. Changing ourselves.

During the implementation of our plan, we will need to focus on monitoring. As we monitor the emissions, in real time, we should be able to ascertain which sectors fail their carbon budgets, given some form of sectoral division of the total budget. This monitoring would need to be a national and transparent exercise, involving all citizens in the same way the financial budget is handled. Actions to reduce the emissions of certain sectors to bring them within their budget would be taken with due regard to burdens and opportunities: burdens will be covered through windfalls and, as far as possible, opportunities generated from the policy steps taken to improve the sector performance.

In Phase One, for example, we know from the LTMS that the largest part of the package in Start Now will be in industrial energy efficiency. Each sector would have its budget, and market instruments such as cap and trade, taxes, tax breaks and incentives, and emissions fines could be set, individually or in combination, to drive industrial energy efficiency – and thus sectoral progress toward the budget.

During the efficiency phase we will have to prepare for the real challenge ahead: the Plateau.

From 2025 to 2030/35, emissions reach a plateau presenting the first major challenge as the budget stays static and (presumably) the economy continues to grow (or perhaps contracts favourably). The required decoupling of growth and emissions that this represents is a very significant challenge. At this point we should be able to 'run the carbon system' in effective parallel with our observations of GDP, job-creation and other major economic indicators. In the Plateau phase we will have our work cut out for us in each group. The static budgets of the energy services and mobility systems (those with the largest budgets) will now be very constrained, and will require major effort, and at times this effort will necessitate either attrition-type activities, or major technological application and opportunity-laden initiatives.

In the emissions decline phase after 2035, our society, our technologies and our economy will look very different. Hopefully our solutions will also promote the 'better life for all'.

What will be needed? I would suggest senior leadership from within the government; a strong basis of data, and scientific and economic analysis based on a robust, credible assessment of abatement potential and costs; stakeholder engagement to enable data collection and cross-sector support, and ongoing processes to build consensus through assessment. One thing we can learn from the LTMS is not to stop once we have got going!

The challenges of such a project are immense on almost every level. We have no or little time. In my view this is one of South Africa's great challenges. Achieving this will be globally significant.

Could South Africa do it?

In answering this question, I am personally pessimistic both for South Africa and also for the global deal. First, it seems so clear that the carbon budget has been blown, both internationally and at home.

> The underlying problem is that so little of the global carbon budget remains. There is no future Scenario – regardless of how the remaining carbon budget is apportioned – in which the South has sufficient space to avoid a decarbonisation transition so rapid that, in anything like a business-as-usual world, it threatens the South's prospects for development. Thus, the only way to secure the earnest engagement of the South is to ensure that it has the assistance necessary to support a decarbonisation transition that

> is rapid and comprehensive, but that nevertheless allows human development to continue unimpeded.[44]

There is little left for the South, little real commitment from the North, and little chance, in my view, that we will achieve the emissions reductions implied by the safety of the 2°C limit. I am also pessimistic that all of us, including the directors and shareholders, the labourers and the consumers of the goods and services, will be able to meet and 'do what is necessary' rather than protect our own interests.

Part of this is that stakeholders ultimately operate with their own interests in mind, and subordinate the broader interests of the environment. This is a lesson we have just learnt again in BP's Deep Water Horizon disaster. BP may, on the surface, be the villain of the peace, but every time we drive a car, we take part in the system that gave rise to that tragedy.

As the Harvard psychologist Daniel Gilbert puts it:[45] 'Global warming is bad, but it doesn't make us feel nauseated or angry or disgraced, and thus we don't feel compelled to rail against it as we do against other momentous threats to our species, such as flag burning.' People tend to have strong emotions about topics such as food and sex, and to create their own moral rules around these emotions, he says. 'Moral emotions are the brain's call to action,' he wrote. 'If climate change were caused by gay sex, or by the practice of eating kittens, millions of protesters would be massing in the streets.'

It is possible that strong evidence-based work with stakeholders will move them from currently-held positions, but in the Climate Change space we are all a stakeholder. To truly succeed we will need a process that includes all of us.

> We don't know how much hotter the planet will become by 2100. But the fact that we face 'only' a 10 per cent chance of a catastrophic 12-degree climb surely does not argue for inaction. It calls for immediate, decisive steps. Most people would pay a substantial share of their wealth – much more, certainly, than the modest cost of a carbon tax – to avoid having someone pull the trigger on a gun pointed at their head with one bullet and nine empty chambers. Yet that's the kind of risk that some people think we should take.[46]

We are on an excursion, in my view, into an unknown ecological place.

---

[44] http://gdrights.org/2009/10/25/a-350-ppm-emergency-pathway-2.

[45] From a 2006 op-ed article in *The Los Angeles Times.*

[46] Robert H. Frank, economist at the Johnson Graduate School of Management at Cornell University, quoted in the *New York Times.*

Chapter 8

Conclusion

'The test of a first-rate intelligence is the ability to hold two opposed ideas in the mind at the same time, and still retain the ability to function. One should, for example, be able to see that things are hopeless and yet be determined to make them otherwise.'

Scott Fitzgerald

The LTMS approach was an exciting way to unlock potential, both in the people who drove the study, as well as in the Options and Scenarios that were produced. It's a great way to fill a 'clean slate' with evidence on mitigation potential and its impacts. It had a surprise factor, an emergent nature, and a highly persuasive impact. It can be taken further. We can learn lessons from the LTMS for the effort we now need to make.

Other country processes

Many countries are also exploring their own mitigation cases, and doing so in their own ways. The differences in approach are quite significant.

The challenge common to all these countries is 'to enable [them] to strengthen delivery of their own development visions and goals through low-carbon, climate-resilient, or "climate compatible" growth strategies. How to address this challenge has been laid out by a growing number of countries in their national plans.'[47] Project Catalyst reports:

> The first generation of these Low Carbon Growth Plans (LCGPs) has shown that many developing, as well as developed countries, are willing and able to commit to ambitious actions on climate compatible growth, based on their own national development priorities and as a contribution to meeting our collective global Climate Change challenge.

[47] Catalyst report.

In their analysis they refer to the stakeholder element in the various processes:

> In most countries development of the plan is taking place within a structured process of stakeholder engagement, to enable:
>
> - Data-collection, analysis and deliberation by different industry sectors and expert institutions and stakeholder groups within the country;
> - Cross-departmental buy-in and coordination within government;
> - Mediation of national stakeholder positions, including identifying and addressing losers; and
> - Broader awareness and public support for change.
>
> However, each country has pursued this engagement with different emphasis, enthusiasm and sequencing between the data analysis, stakeholder engagement and policy formation phases. In some countries, such as India and Bangladesh, criticism of the initial level of stakeholder engagement from local and national stakeholders has precipitated further rounds of consultation.
>
> The South African process stands out for its stakeholder engagement, integrated throughout the strategy development process. In particular the research base fed into a facilitated stakeholder process. Central to the process was the Scenario Building Team (SBT), which brought together strategic thinkers from key sectors across government, business and civil society. The SBT gave detailed comments on the assumptions and data used by the research teams and its thinking and dialogue was advanced by the research commissioned. In particular the coordinator reports that the team was shocked that the gap between the 'Growth without Constraints' and 'Required by Science' Scenario was so large, and this caused them to change their approach to thinking about possible futures. The Scenarios document agreed by the SBT was opened to consultation with a broader set of stakeholders, including CEOs and representatives from NGOs and labour as well as ministers in government.
>
> In South Africa the facilitated stakeholder process was critical to building consensus around the results and rigour of the research methodology, and building up a broad base of support for action.

As Harald Winkler, LTMS project leader, relates: 'The creation

> of the Scenario Building Team in itself is an important outcome. Results shaped and endorsed by a set of strategic thinkers from a diversity of stakeholders carry much greater weight that a simple research report. This team of people has the potential to continue playing an important role in future.' However, South Africa's process has taken a long time, and may not be a model suited to the political culture of other countries.[48]

Harald and I have been invited by a number of countries (Brazil, Peru, Zambia, Colombia, Ghana, and Chile with others possibly following) to work with them in exploring an 'LTMS+' approach. I look forward to learning how these countries will continue to address the combination of evidence and stakeholder participation, the socially accountable and the research driven approach, within a context of different cultures and histories. I am also enthused by the potential of this 'South-South' collaboration. Learning from each other is essential.

'South Africans as a whole are much more familiar with Scenario processes than perhaps anywhere else in the world. The person in the street is in fact familiar with such processes. These thought experiments are not something new to South Africans as a result.'
Bob Scholes, CSIR

An LTAS?

The idea of an LTAS (Long-Term Adaptation Scenarios) for South Africa was mooted during the LTMS process. The South African government and some of the leading South African climate impacts scientists are supportive of the idea, but thus far resource availability has constrained immediate implementation. Further discussion is needed to refine the idea, however, and we're hopeful that an LTAS may well be launched in South Africa shortly.

'If we look at the people that undertook the work and under what auspices it was undertaken, it had a lot of credibility. It was not an independent study commissioned by a consultant.'
Kevin Nassiep, CEF

'Structuring an LTAS would represent a very difficult problem: it lacks the single coherent objective of the LTMS. It would be possible, and should be motivated for in the form of an integrated assessment for South Africa. Such studies exist world-wide but at a resolution that is not helpful at national level. Technically it is something that could be done but it would take several years to set up. The Adaptation material would, by its nature, be less conflictual, fortunately. There are not such strong trade-off issues.'
Bob Scholes, CSIR

[48] Ibid. p 29.

An LTAS would be a complex process. Some thoughts on how an LTAS might work reveal its fascinating potential.

The LTAS would, in every sense, be the sister process to the LTMS: the one revealing the cost, economy-wide impacts and 'climate-effectiveness' of a series of emission and development pathways (the cost of action); the other similarly revealing the cost, economy-wide impacts, and climate impact of a series of differing climate realities. Having the results of both processes would enable some comparison between these two costs, as a motivator to act.

An LTAS may borrow the overriding principles from the LTMS:

- The engagement of stakeholders in both agreeing to the inputs to the research (where appropriate) and to the assessment of the emerging Scenarios and their impacts on stakeholders;
- The production of Scenarios (this could be a complex matrix between various CO_2 concentrations and the various ranges of modelled impacts);
- The inclusion of top-quality science; and
- An economy-wide assessment of the Scenarios.

The advantage of a granular assessment (knowing, for example, the specific climate impacts in defined areas of the country will help planning immensely in respect of infrastructure, economic activity and so on) is of course dependent on the ability of science to drill down to the level of fine data. To put this practically: can our science tell us what is likely to be the impact under a 500 ppm Scenario for a specific 25 square kilometre area in 2030? And if so, can we with some accuracy count the cost and attribute it in a way that has practical value? Within each of these grids, one can assess the assets (human, infrastructural and natural) that can be impacted by future Climate Change, if one has the data.

Helpful as this may be in theory, the interesting additional question is what the role of stakeholders in such a process would be. Why not just get the best scientists to model the impacts and costs and present this to the country? After all, the levels of engagement and conflict will not reach the level it had in the mitigation debate.

Once again this raises the question of the legitimacy and credibility of assessments: put plainly, people do not believe the extent of expected Climate Change and other shocks, and seem content to believe that life will go on more or less unchanged. A report, even backed with impeccable science (if that were possible) may still be pooh-poohed if it describes what people see as 'unlikely' and 'inconvenient'. The science of the future is also built on models, and is thus constrained by their uncertainties. We still have to learn how to restate the risks in a way that encourages action at scale.

The value of stakeholders traversing the same ground with the scientists, as was shown in the LTMS, is that people then do not argue with the science – since they see themselves as being part of the process of generating, or at least understanding it. In an LTAS, however, stakeholder identification would be extremely difficult!

Stakeholders within each granular study area would have to be identified (many will, of course, have influence over multiple areas). The basic principle is: if the models can show us the projected impacts on each area, stakeholders can then engage in the quantification of the remedial, preventative and other adaptive actions that are required for that area. These can be economically and physically described and plans of action can be built on them.

Scenarios would come into play depending on the success or otherwise of the world's emission reduction efforts. Scenarios could (depending on the ability of science to deliver this) present impact pictures for the study areas at different levels of carbon concentration in the atmosphere (e.g. 400 ppm, 450 ppm, 500 ppm or 600 ppm) as and when appropriate. In this way we could see the limits of adaptation.

The overall value of such a process may be significant in many senses: popularisation of the knowledge of what may lie ahead for us, planning and preventative activity, climate-proofing infrastructure investment, and so on. But perhaps the greatest benefit will be the impetus it will give to the mitigation effort, both locally and internationally.

The government will see from the study what the costs of adaptation are likely to be under different emissions Scenarios, and will be able to compare costs of adaptation with costs of its own mitigation and the effort of ensuring a global deal is effective and workable. In short, as a policy tool this is very important. In addition to the government, participating, stakeholders will, for the first time, see (through their engagement in the process) how the Climate Science works, how they will be impacted, and thus will be involved in creating public trust in and knowledge of these results, including why they may have to bear up to significant mitigation costs, such as more expensive electricity. This will powerfully motivate mitigation action by these stakeholders (as key emitters) as well. In fact, it is likely that the LTAS would have a stronger mobilisation effect than the LTMS.

Having said all this, an LTAS would need careful thinking and consideration, and may be far more difficult to achieve than the LTMS.

'Adaptation is a much tougher nut to crack, process-wise, than the LTMS, as the range of stakeholders is vast, and the impacts extend across many more sectors than is the case in mitigation. The science of adaptation is also a lot less certain, firstly because climate Scenarios that drive the need for adaptation are still evolving. It is also far more difficult to identify how an adaptation response to Climate Change differs from everyday responses to climate variability, and other risks that may not be climate-related. Mitigation actions, by contrast, are fairly easy to encapsulate and describe. Then there is the thorny issue of economic diversification as an adaptation response to global shifts in demand for energy sources and other mitigation responses. The issue is at the moment dubiously wrapped up with adaptation to adverse impacts of Climate Change in the UNFCCC negotiation forums, but is really a mitigation issue. An LTAS would therefore have to be very carefully scoped out.'
Guy Midgley, director SA National Botanical Society

Some personal reflections

The LTMS and the approach the South African government took in response to the study (and continues to take in the negotiations) has been praised internationally. I am proud of our work, and of the team that produced it. The challenge, however, is to turn the creative burst of the LTMS into action. This strategy development, and the policy package that goes with it, may soon see many of the SBT players back at work again. I can only hope that South Africans continue to pursue a solution to address Climate Change and drive a new low or no-carbon economy that also assists development and poverty reduction. I hope we can find ways to carefully dismantle, over time, the carbon-based economy. I hope that those I came to know at the coalface of the emissions systems know that this is what we have to do, eventually, and that we need to start doing it now.

It is a great help to start with South African stakeholders on the same page. Although the consensus that built around the LTMS was essentially a dream of sorts, it was and is a brave dream, worthy of realisation. South Africans can, when required, roll up their sleeves. I watched them do it.

My country has a wonderful history of social accountability, and despite its problems, still does not shy away from airing proposed decisions in public and letting the voice of the people be heard. But it will always be difficult to structure that voice in a way that really informs decisions, or opens up new vistas. People are by nature self-serving, and nowhere is that more true than in the view that we have a right to cheap and abundant electricity or fuel supply, regardless of the damage this may be causing. Our challenge is to structure the dialogue in such a way that we harness our imaginations and then inform current action. The LTMS was, in many senses, an experiment in process management aimed at achieving this new form of conversation – an attempt, and a good one at that, to try something different.

Stakeholder-driven processes are often useful as formal or informal tools of accountability. Accountability, of course, is a concept in ethics with several meanings. With the issue of Climate Change it gains a new, intergenerational ethical context. Accountability is answerability to these new ethical constructs. The LTMS left these ethical issues to the implicit space, but through an exposure to evidence, arrived at ideas essentially aligned to them.

By taking the risk of letting people take the time to develop the data behind the ethic, the evidence behind the agreement, and the creativity to imagine futures, the South African government started the country's work on an audacious footing, but one that will pay off, I have no doubt.

Behind the LTMS lies our human dilemma. We are deep into the age of consequences. We have a simple option: face the consequences of our impact on the ecosystem or embrace the collective activism that drives new, imagined solutions.

Annexure 1

Some climate facts pertinent to South Africa

The World Resources Institute's[49] work in 2005 gives us a look at data relevant at that time. The report correctly points out that before any planning work can be undertaken one needs to have accurate emissions data. There is either a lot of data (which makes the job of using the data practically difficult) or too little data (which ups the use of assumptions, and may compromise results). Data may also range in its reliability (these data challenges confronted us many times during the LTMS process).

Internationally: the concentration of CO_2 in the atmosphere is still rising exponentially and is approaching 400 ppm; world emissions add around 2 ppm per annum.[50]

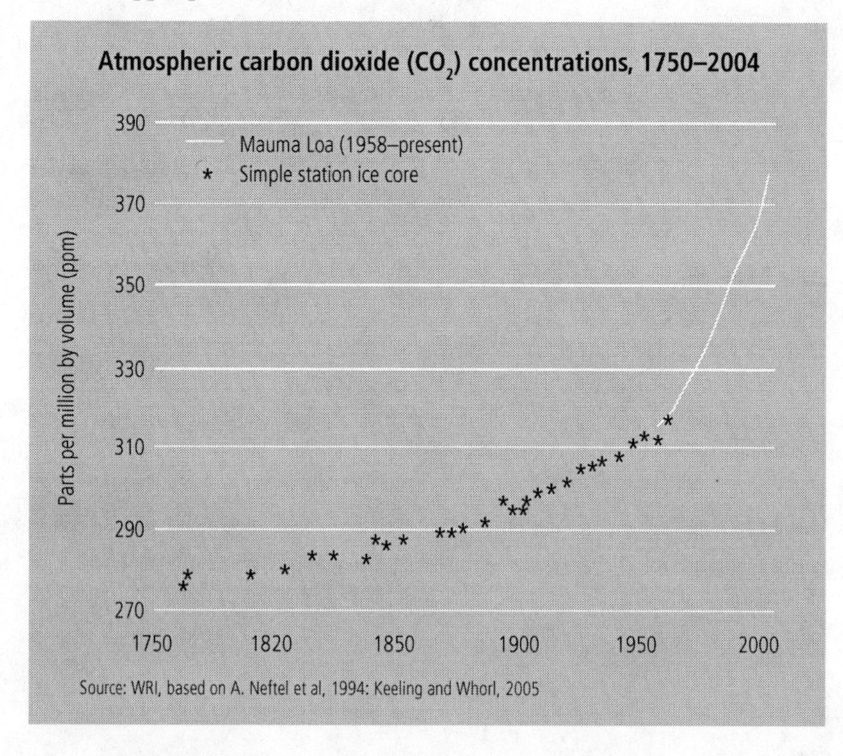

49 *Navigating the Numbers*: WRI; Baumert, Herzog, Pershing 2005, see www.wri.org.

50 Ibid.

This emissions increase is, in large part, directly attributable to human activity:

> In the past 200 years, more than 2.3 trillion tons of CO_2 have been released into the atmosphere due to human activities relating to fossil fuel consumption and land-use changes. Fifty per cent of these emissions have been released in the 30-year period from 1974 to 2004. The largest absolute increase in CO_2 emissions occurred in 2004, when more than 28 billion tons of CO_2 were added to the atmosphere from fossil fuel combustion alone. The year 2004 also represented the largest percentage increase in emissions since 1976.[51]

There are six primary greenhouse gases, CO_2 being the main culprit (around 77% of all GHG emissions), followed by methane (14%) and nitrous oxide (8%), and then some other really esoteric ones; sulphur hexafluoride, the perfluorocarbons and hydrofluorocarbons make up the small balance. By 2005 the total 'supply' tonnage of GHG emissions from fossil fuels was approaching 30 billion tonnes per annum. The total CO_2 equivalent emissions in that year were around 450 gigatonnes. That is a lot of elephants floating in the sky.

As far as the historical stock is concerned, the position would be considerably different, in favour of developing countries: China now emits more than the US in supply, but will only 'catch up' on stock in some 20 or so years, on a business as usual approach.

In South Africa, the source of emissions is profiled quite differently from say Brazil, but is more similar to a country such as China. Eskom, the national utility, contributes some 220 of the total 430 Mt; Sasol (the coal and gas to oil operation) some 72 Mt. In 2009 the emissions figures were published for comment.[52] Preliminary figures for 1999/2000 reveal the following.

---

51 Ibid.

52 SA *Government Gazette* 32490 of 12 August 2009.

Sector transmission trends and percentage changes from 1990								
Sector	GHG emissions CO_2eGg							
	1990 (CO_2e Gg)	% of total	1994 (CO_2e Gg)	% of total	2000 (CO_2e Gg)	% of total	2000 % change from 1994	2000 % change from 1990
Energy	260 886	75.1	297 564	78.3	344 106	78.9	15.6	31.9
Industrial processes and product use	30 792	8.9	30 386	8.0	61 469	14.1	102.3	99.6
Agriculture	40 474	11.6	35 462	9.3	21 289	4.9	–40.0	–47.4
Waste	15 194	4.4	16 430	4.3	9 393	2.1	–42.8	–38.2
Total (without LULUCF)	**347 346**		**379 842**		**436 257**		**14.8**	**25.6**

The trends for individual gases gave a different picture. There was a uniform increase in emissions for all types of greenhouse gases, with no decreases. Methane showed the highest increase, recording an increase of more than 76% from 1990 to 2000. Nitrous oxide showed the lowest increase from 1990 to 2000, of 2.7%.

Gas emission trends and percentage changes from 1990									
GHG emissions CO_2eGg	1990 (Gg)	% of total	1994 (Gg)	% of total	2000 (Gg)	% of total	1994 % change from 1990	2000 % change from 1994	2000 % change from 1990
CO_2	280 932	80.9	315 957	83.2	353 643	81.1	12.5	11.9	18.6
CH_4	2 053	12.4	2 0527	11.4	3 624	17.2	0.2	76.2	76.5
N_2O	75	6.7	67	5.4	76.7	1.3	–10.7	14.5	2.7
CF_4	–		–	–	0.303	0.5		–	–
C_2F_4	–		–	–	0.027	0.06		–	–
Totals CO_2eqGg (without LULUCF)	**347 346**		**379 842**		**436 257**		**9.4**	**14.8**	**25.6**

This sums up in broad terms where South Africa stood in terms of its emissions. More importantly, before 2005 no extensive work had been done to understand the consequences of actively reducing these emissions.

Annexure 2

The terms of reference for the LTMS

Climate Change is a major threat to our planet. We are contributing to Climate Change by emitting greenhouse gases into the atmosphere. Reducing these greenhouse gases significantly is urgently required. This action is called mitigation. Currently a number of industrialised nations have taken mitigation targets under the Kyoto Protocol, during a first agreed commitment period, which ends in 2012. South Africa and other developing nations do not need to act and do not have such targets – for now.

Given that most greenhouse gases are emitted in the production of energy, mitigation has significant effects on economies, and is central to any countries development path.

Over the next number of years, South Africa, committed as it is to the multilateral process under the UNFCCC and its Kyoto Protocol, will be required to engage with the negotiations around developing country actions. The decisions taken at COP11 and COP-MOP1 in Montreal in December 2005 make clear that this will proceed initially in two tracks – further commitments for Annex I Parties under Article 3.9 of the Kyoto Protocol, and a dialogue on long-term cooperative action under the Convention. Beyond the two years of the Convention process, South Africa needs to define what its position on future commitments for the more rapidly industrialising developing countries.

In addition to mitigation, South Africa must also concentrate on the appropriate actions it should take to adapt to the impacts of already occurring and predicted Climate Change. This action is called adaptation. Whilst one element of the future negotiations will be focused on South Africa's challenge to adapt to current and projected impacts, adaptation to impacts cannot occur while no mitigation is taking place, as this assumes an indefinite adaptation to an indefinitely increasing problem. Hence there will come a point where it is no longer possible to adapt our way out of the projected impacts. It is clear that South Africa will, in any event, have to spend money and resources on adaptation, as impacts of Climate Change caused by global emissions will be felt with increasing intensity over the next 50 years and more. In this process, the costs of adaptation are treated as 'base-line costs', which will be incurred in every Scenario. Our own mitigation efforts will be unlikely to affect this need to adapt. But mitigation has to be a multi-lateral effort, and our contribution to world greenhouse gas emissions, whilst small, is significant in a number of aspects. First is the fact that we are probably the highest emitter internationally per capita per GDP. Second, we account for almost half of Africa's emissions. Third, we have taken a leadership position in the UNFCCC process, and must lead by example. Fourth, it is likely that getting the USA on board will require commitments from major emitting countries in the developing world. If South Africa does not act to reduce GHG emissions, others are less likely to do so and eventually our common future is threatened. The impacts of Climate Change will be felt most in poorer countries and communities. Finally, and most significantly, there is a moral imperative to act as one world, as science tells us that the Climate Change problem is much greater than first described.

It is necessary to tackle the root cause of the problem, by reducing greenhouse gas emissions. Given its coal-dominated energy economy and emissions profile, the sharing of mitigation costs internationally has critical strategic implications for South Africa. The major focus of this process is on defining Long-Term Mitigation Scenarios.

South Africa has the opportunity to proactively define approaches and development paths that we – as a society – consider desirable. The South African delegation has already played a constructive role in taking the multi-lateral negotiations forward. *What is needed is a national process of building Scenarios of possible futures, informed by the best available research and information.* This will help South Africa to define not only its position on future commitments, but also shape its climate policy for the longer-term future. Cabinet has given a mandate for such a process to be undertaken and the Department of Environmental Affairs and Tourism (DEAT) has asked the Energy Research Centre (ERC) to project manage this process. The proposal by the ERC outlines the working method, proposed approach, time line, and cost developing a Scenario Building process informed by the best available research.

The objective and proposed outcomes of the envisaged process were set out as follows:

1. South African stakeholders understand and are focused on a range of ambitious but realistic Scenarios of future climate action both for themselves and for the country, based on best available information, notably Long-Term Emissions Scenarios and their cost implications;
2. The South Arican delegation is well-prepared with clear positions for post-2012 dialogue;
3. Cabinet can approve (a) a long-term climate policy and (b) positions for the Convention dialogue up to COP-13; and

4. Cabinet policy based on the Scenarios will assist future work to build public awareness and support for government initiatives.

The LTMS public brochure

Introduction

Climate Change is one of the greatest threats to our planet and to our people. South Africa is especially vulnerable to the impacts of Climate Change. At the same time South Africa emits high quantities of the greenhouse gases, which are causing Climate Change: in fact our country is one of the highest emitters per capita per GDP in the world. We are helping to cause the problem and we are also its victims.

Reducing emissions of greenhouse gases is called mitigation. Responding to the impacts of Climate Change is called adaptation. The proposed Long-Term Mitigation Scenario process refers primarily to mitigation activities. A certain amount of adaptation will be necessary, no matter what we do. But it is also true that there will come a point where it will not be possible to adapt our way out of the problem.

South Africa is an active participant in the international process of combating Climate Change and regulating the emissions of greenhouse gases. We are signatories to the United Nations Framework Convention on Climate Change as well as the Kyoto Protocol. We take the issue of Climate Change very seriously and have shown world leadership in the UN negotiations. Our actions must also speak as loudly as our words in the negotiations: we need to show leadership by example. This we can do by preparing a course of action for our country.

Under the Kyoto Protocol, at least until 2012, we, together with most developing countries, have no binding greenhouse gas mitigation obligations. However, this is likely to change some time after 2012 and means that at some point South African will be required to start cutting its emissions. South Africa is in fact already formulating plans to reduce GHG emissions.

Over the next few years in the negotiations, South Africa will be required to engage deeply with the issue of mitigation obligations. We will need to be ready and prepared, armed with a detailed plan and sets of negotiation positions. This plan will have to contribute to the international effort to lower emissions while meeting the development needs, especially of our poorer communities. We need to connect energy needs, mitigation plans, and policies such as the Accelerated and Shared Growth Initiative.

We need to accurately determine the costs, benefits, and opportunities for mitigation activities. We will choose a time horizon of both 25 and 50 years, which are fair time frames for medium- and long-term planning when we speak of power generation, as well as for other emission sources such as from industry, transport and housing.

Scenario planning

Mitigation is a delicate balance between development needs, available technology, cost to the economy, and policy intervention. South Africa has the opportunity to proactively define approaches and development paths that we – as a society – consider desirable. We cannot, for example, agree to a mitigation target that we cannot afford and will not reach. At the same time, there is a huge opportunity for international investment in climate-friendly technology, which can help us grow more and create new industries. In other words, we need to work out a range of paths that work for our country. This includes all major emitters: our electricity utility, our private sector, and our public sector.

In order to determine these paths, our Cabinet has decided that what is needed is a national process of building Scenarios of possible futures, informed by the best available research and information. This will help South Africa to define not only its position on future commitments under international treaties, but also shape its climate policy for the longer-term future.

Stakeholders from government, business and civil society agreed at the National Conference in October to embark on this process, seeking to protect the climate while meeting the development challenges of poverty alleviation and job creation.

Scenario planning has already been influential in our history and has proved its ability as a process to shape policy and other choices. A Scenario is *a structured account of a possible future*. A Scenario describes a future that *could* be rather than one that *will* be. A group of Scenarios are *alternative dynamic stories* that capture key ingredients of uncertainties of the future. They reveal the *implications of current trajectories*, thus illuminating *options for action.* These options for action are then presented to government in order to assist it in making the correct policy choices.

The method

From May 2005, and for a period of about 18 months, we will put together a Scenario Building Team that will develop the Scenarios. The Team will be made up of directly interested stakeholders from the countries major emitters, from government, as well as from other interested parties. A careful process of stakeholder selection will ensure that the Team contains the correct people for the task. Expert independent process facilitators will facilitate the team with international experience in Scenario Building and Climate Change issues. Four Research Units, covering Energy Emissions, Non-Energy Emissions, Macro-Economic Modelling, and Climate Change Impacts, will support the Team. These support Units will contain our leading researchers.

The Scenario Building Team will start building the Scenarios based on research information and internal data later in 2006. The Team will build its own rules, which are likely to include strict principles regarding confidentiality and protection of proprietary information.

The final report of the Team will be made public.

Who?

The Department of Environmental Affairs and Tourism (DEAT) as the focal point for Climate Change in South Africa will convene and manage the process, which will be overseen by an inter-ministerial group. DEAT has appointed the Energy Research Centre at the University of Cape Town (ERC) to project manage the entire process. The ERC is undertaking the task of convening and contracting the process specialists and ensuring their independence. Similarly it is setting up the personnel of each of the four Research Support Units.

Outputs

The desired objectives of the proposed process are:

- South African stakeholders understand and are focused on a range of ambitious but realistic Scenarios of future climate action both for themselves and for the country, based on best available information, notably long-term emissions Scenarios and their cost implications;
- The SA delegation is well-prepared with clear positions for post-2012 dialogue;
- Cabinet can approve (a) a long-term climate policy and (b) positions for the dialogue under the United Nations Framework Convention on Climate Change; and
- Cabinet policy based on the Scenarios will assist future work to build public awareness and support for government initiatives.

Annexure 3

Scenario examples

Clem Sunter, in his Wits Business School lecture of 16 Jan 2008,[53] explains how the approach works (applied to another Climate Change process):

> The two variables selected to be the axes for the Climate Change Scenario gameboard are the level of political will and the presence or absence of international co-operation. Combining these variables gives four possible Scenarios, which were formulated by a group of policy experts recently in London. They outline possible futures after the Kyoto Agreement comes to an end in 2012.

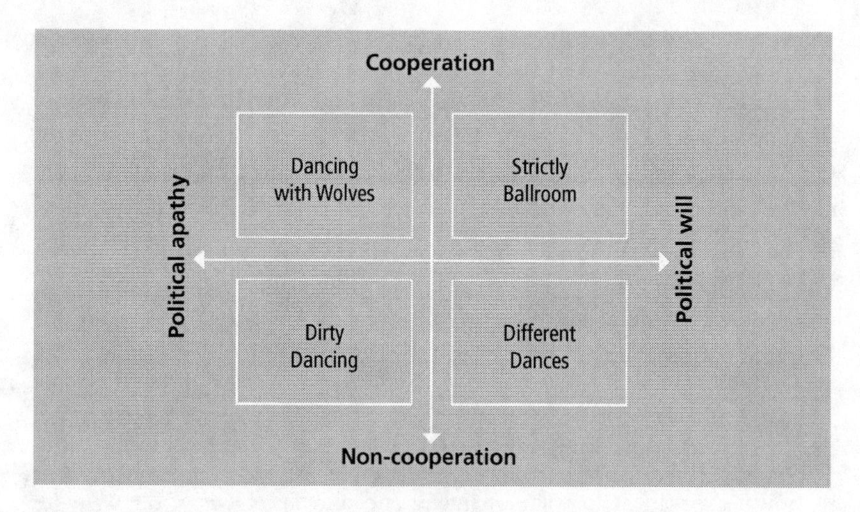

> **Dirty Dancing.** In this Scenario, no new agreement of substance replaces Kyoto. Everybody goes their own way, doing nothing to curb carbon emissions. It's all about the economy, stupid! Energy security, resource nationalism and protectionism rear their heads; and Climate Change dissidents wage a successful war with the scientific community to sow confusion in the public mind.
>
> **Different Dances.** In this Scenario, some countries (like those in Scandinavia) and some states in America (like California where

[53] www.mindofafox.com/casestudies.php.

Arnold Schwarzenegger is now dubbed the 'Greenerator') come to the party with carbon emission reduction programmes (including successful carbon trading schemes). There is no coordination between these programmes so they have limited impact, except to provide a role model for other countries to follow. Nevertheless, it's better than nothing.

Dances with Wolves. This Scenario unveils a formal agreement with much fanfare but no teeth. Immediately after it is signed, countries – called 'the wolves' – start cheating on the agreement. There is no retribution for their misbehaviour because the agreement is vaguely worded and full of loopholes.

Strictly Ballroom. This is the virtuous Scenario where countries sign up to an agreement containing clear objectives and targets and keep to it. They waltz together in a co-ordinated programme of action, sharing best practice and new technologies. The US leads the way with China and India in hot pursuit. A global carbon trading market is established.

These Scenarios are arrived at using a 'play', where the participants agree to the Scope of the game, the Players, and Rules of the Game.

Holmgren's work is also fascinating, and he uses the quadrant approach as well (an approach that necessarily leads on to 4 Scenarios): in his example,[54] increasing Climate Change and increasing scarcity of oil make up the axes.

He develops four stories, which fit into the quadrants, which he calls 'Energy Descent' Scenarios. So if energy decline rates are slow (as a result of a failure to mitigate + availability of fossil resources) and Climate Change is correspondingly severe, then brown-tech is the name of the game, with centrist, top down constriction by the State, and severe impacts and human disruption.

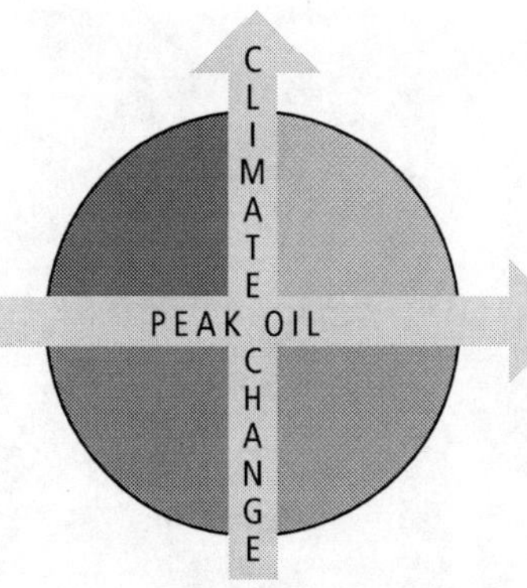

54 Holmgren, www.mindofafox.com/casestudies.ph, p 60ff.

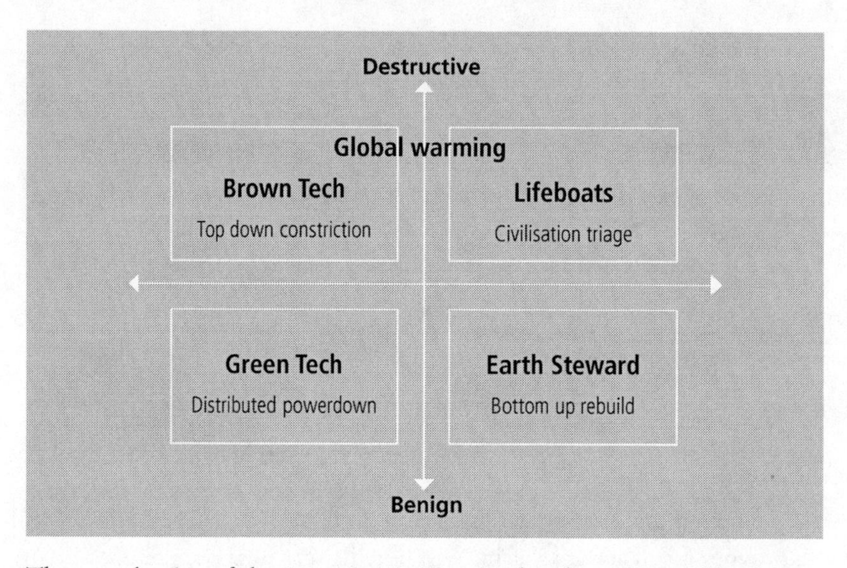

The most benign of the Scenarios is GreenTech, where the world can keep pace with a mild rate of Climate Change and a slow oil decline. The toughest Scenario is of course Lifeboats, the McCarthy world of 'The Road'.

The consistency of the stories is the trick: they 'make sense' against the unfolding of the various combinations of variables unfold. Happily or sadly, they could all be true.

The consistency can be illustrated by using common descriptors for the Scenarios, which in Holmgren's example look like this:

Scenario	Energy and agriculture	Settlement form and mobility	Economy and money	Politics	Gender	Culture and spirituality
Brown Tech Top down constriction	Centralised power High tech efficiency Non conventional oil, gas, coal, nuclear Bio shelter agriculture	High density cities Electric private transport Hinterland abandonment Mass migration	National banks and currencies	Nationalist/ fascist Class structure and rights Price rationing Pop control	Male dominated and blended	Super rationalist/ fundamentalist dichotomy
Green Tech Distributed powerdown	Distributed network Conservation Gas, wind, solar Forest, organic agriculture	Compact towns and small cities Electric public transport Telecommuting	Regional currencies and funds	City state and hinterland Markets/ rationing Democracy?	Balanced and blended	Humanist/ Eco-rationalist

Scenario	Energy and agriculture	Settlement form and mobility	Economy and money	Politics	Gender	Culture and spirituality
Earth Steward Bottom up rebuild	Distributed local hydro, methane Industrial salvage Forest, organic and garden agriculture	Ruralisation of suburbia Rural resettlement Minimal mobility	Local currency barter	Town and bioregion Participatory democracy? Neo-feudalism	Female dominated and gendered	Earth spirituality
Lifeboats Civilisation triage	Distributed local Forest rangeland Industrial salvage Oasis agriculture	Hamlet and gated communities Nomad	Household and barter, precious metals	Feudal system Patriarchal authority	Male dominated and gendered	Warrior cult

Annexure 4

Methodology for Scenario building

There are a number of steps that one can use to build these Scenarios. Here is one approach:

Step 1: Identify the focal issue and the horizon

In general it is advisable to start with a specific decision or question, then building out towards the environment. In the LTMS, this question was already set: what mitigation actions could South Africa undertake? A narrow focus will prevent the Scenarios from drifting into broad generalisations about the future of society or the global economy. We had this narrow focus.

When determining the focal issue it is important to consider the appropriate time-horizon for the Scenarios, because it will affect the range of issues to be considered within the Scenario development process. In the LTMS we had decided on a time horizon of 2050.

Step 2: Identification and analysis of the drivers

The next step is usually to identify the key drivers that will influence the listed key forces at macro- and micro-level. These are Holmgren's variables. The first stage is to examine the results of environmental analysis to determine which are the most important factors that will decide the nature of the future environment within which the organisation/country/world operates. These drivers can also be called 'variables' (because they will vary over the time being investigated, though the terminology may confuse scientists who use it in a more rigorous manner). Users tend to prefer the term 'drivers' (for change), since this terminology is not laden with quasi-scientific connotations and reinforces the participant's commitment to search for those forces that will act to change the future. In LTMS we referred to 'drivers' and 'assumptions'.

In this way a conceptual model of the relevant environment is built, and includes critical trends and forces, and maps out the cause-and-effect relationship among the forces.

Step 3: Rank by importance and uncertainties

The drivers can then be ranked using factors such as most important to least important, most to least uncertain.

Step 4: Selecting Scenario logics

In the classic Scenario play, the drivers are then ranked and finally the two principal drivers set up in the Scenario logics, namely the axes. The focus of attention should be on the 'high important/low uncertainty' and on the 'high important/high uncertainty' quadrants of the matrix. Determining the axes of the Scenarios is the crucial step in the entire Scenario-generating process. This is also the step in which intuition, insight, and creativity play their greatest role.

The main goal (and challenge) is to end up with just a few Scenarios that make a difference for the decision-maker.

Step 5: Fleshing out the Scenarios

The different Scenarios are essentially internally consistent story lines for each quadrant. The stories will be characterised by:

1. **Plausibility**: The selected Scenarios must be plausible, this means that they must fall within the limits of what might conceivably happen.
2. **Differentiation**: they should be structurally different, meaning that they should not be so close to one another that they become simply variations of a base case.
3. **Consistency**: They must be internally consistent. The combination of logics in a Scenario must not have any built-in inconsistency that would undermine the credibility of the Scenario.
4. **Decision-making utility**: Each Scenario, and all Scenarios as a set, should contribute specific insights into the future that will allow on the decision focus that was selected.
5. **Challenge**: the Scenarios should challenge conventional wisdom about the future.

Once the Scenarios have been fully developed, they have to be elaborated. There are many ways to elaborate the description of Scenarios, but there are three very important features:

1. The Scenarios are named: a highly descriptive title is usually used: short enough to be memorable; descriptive enough to transmit the essence of what is happening in the Scenario.
2. The Scenarios contain compelling 'story-lines': Scenarios are narratives of how events might unfold between the present and the selected time-horizon; each should provide the dynamics (logics) assigned to it. In simple terms the Scenario should tell a story that should be remarkable, convincing, logical, and plausible.

3. They are supported by a table of comparative descriptions: this provides planners and decision makers with a sort of 'line item' description that details what might happen to each key trend or factor in each Scenario (see the Holmgren table in Annexure 3).

Step 6: Implications of Scenarios

This is the stage at which the Scenarios are turned back on the initial question, and used to evaluate or develop strategies.

Annexure 5

Assessment examples

The authors refer to the following Assessments as examples of the format:

Millennium Ecosystem Assessment

The Millennium Ecosystem Assessment, released in 2005, assessed the consequences of ecosystem change for human well-being. The MA consisted of a global assessment and 34 sub-global assessments to assess current knowledge on the consequences of ecosystem change for people. The MA brought about a new approach to the assessment of ecosystems: a consensus of a large body of social and natural scientists; the focus on ecosystem services and their link to human well-being and development; and identification of emergent findings. The MA findings highlight the strain that human actions are placing on the rapidly depleting ecosystem services but also that appropriate action through policy and practice is possible. (www.MAweb.org)

International Assessment of Agricultural Science and Technology for Development

The International Assessment of Agriculture Science and Technology for Development (IAASTD), released in 2008, was an intergovernmental process that evaluated the relevance, quality, and effectiveness of Agricultural Knowledge, Science, and Technology (AKST) and the effectiveness of public- and private-sector policies, as well as institutional arrangements in relation to AKST. The IAASTD consisted of a global assessment and five sub-global assessments using the same assessment framework, focusing on how hunger and poverty can be reduced while improving rural livelihoods and facilitating equitable, environmental, social, and economical sustainable development through different generations and increasing access to and use of agricultural knowledge, science, and technology. (www.agassessment.org)

Intergovernmental Panel on Climate Change

The Intergovernmental Panel on Climate Change (IPCC) released its fourth report (AR4) in 2007. The IPCC was established to provide decision makers with an objective source about Climate Change. Similar to the MA, the IPCC does not conduct any research or monitor specific data and parameters; it assesses the latest scientific, technical, and socio-economic literature in an objective, open, and transparent manner. Ecosystem services are addressed in the fourth report of the IPCC by the reports of Working Group II (Impacts, Adaptation and Vulnerability) and Working Group III (Mitigation of Climate Change). The findings of AR4 highlighted a number of overarching key issues in relation to ecosystems and the services they provide for climate mitigation and adaptation. Specifically, the report drew links between the loss of ecosystem services and the reduction of societal option for adaptation responses. (www.ipcc.ch)

Land Degradation Assessment of Drylands

The Land Degradation Assessment of Drylands (LADA) is an ongoing assessment that aims to assess causes, status, and impact of land degradation in drylands in order to improve decision making for sustainable development at local, national, sub-regional, and global levels. Currently the LADA is focusing on developing tools and identifying available data that will be required to discover status and trends, hotspots of degradation, and bright spots (where degradation has been slowed or reversed). (http://lada.virtualcentre.org/)

Global Environment Outlook

The Global Environment Outlook (GEO) is the United Nations Environment Programme's ongoing assessment of the environment globally. The fourth GEO was released in 2007 and consists of a global assessment and sub-global assessments. GEO-4 provides information for decision makers on environment, development, and human well-being. (www.unep.org/geo)

Annexure 6

Assessment methodology

Once the Assessment process actually begins, the framework for the approach must be set:

> Central to the coherence of an assessment is the design or adoption and use of a conceptual framework – a common understanding of what the assessment aims to do. A single agreed conceptual framework guides the assessment, allowing multiple practitioners to work within the same boundaries and understanding of what is being assessed, and therefore allows integration of the components of the assessment. Conceptual frameworks help clarify the complex relationships between elements of the human–ecological system, including how those relationships may be changing over time.

In the case of the LTMS, this framework was the Scenario approach already described. This is what the authors say of this component:

> The Millennium Assessment contained three basic components: assessment of condition and trends, Scenarios, and responses. Although each of these can be undertaken by a separate working group of scientific experts, at many of the sub-global scales of the MA all three components were addressed by the same group. In many cases it would be better to undertake these three components sequentially, so as to first assess the condition and trends of ecosystems, their services, and human well-being; second, develop Scenarios of change; and third, assess available and potential responses. In practice, however, these elements also need to interact closely, and what is learned from doing any one of these components can inform other technical parts of the assessment. The most important issue is that these components remain connected – through use of a joint conceptual framework, terminology, and approach.

Of the Scenario element:

> The Scenarios component should aim to develop a set of Scenarios providing descriptive story-lines, supported by quantitative

approaches, to illustrate the consequences of various plausible changes in drivers, ecosystems, their services, and human well-being. Scenarios are not attempts to forecast the future but rather are designed to provide decision makers with a better understanding of the potential consequences of decisions. They also help to understand the uncertainty about the future in a creative way and can help explore new possibilities to respond to change.

An adapted diagram below shows the various stages of Assessments:

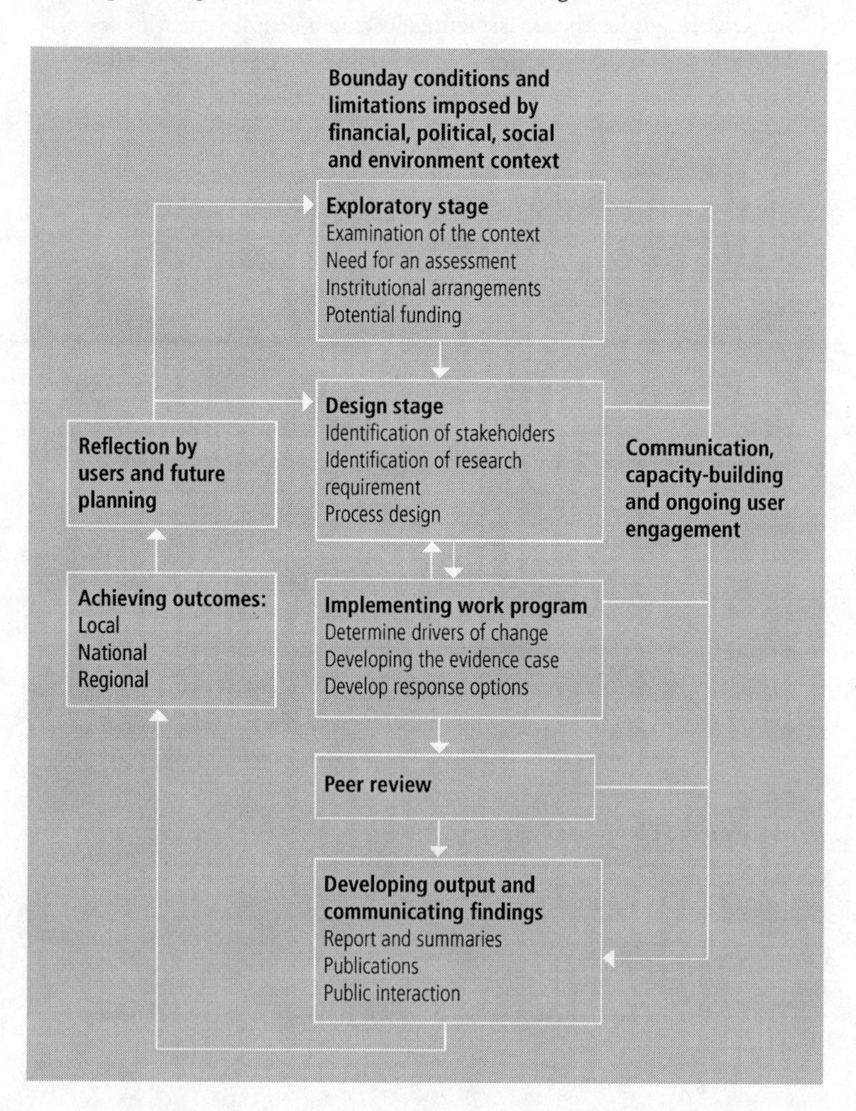

Annexure 7

Management in the LTMS

Obviously the first step before embarking on a large project is to get the management ducks in a row. This could be seen as a tedious element of the process, but actually the way it is structured can bring a great deal of creativity to the process. Reid *et al.* add the following, as far as Assessments go:

> The governance (including leadership) of an assessment can be a critical factor in ensuring user engagement, raising funds, and overseeing progress in implementation of the assessment. It is also crucial to ensure legitimacy and credibility. A model that has been found to be effective in the Millennium Assessment and other assessments is having the assessment overseen by a technical Steering Committee or Assessment Panel and an associated 'User' Advisory Committee or Assessment Board. In some cases, involving both the technical experts and users in a single committee might work well; in other instances, it may be more appropriate to establish a separate Advisory Group to represent the various users.
>
> Typical functions of a Steering Committee would be to:
>
> - Promote coordination among the institutions and individuals carrying out the assessment;
> - Develop the detailed assessment design (what information will be produced by which individuals and institutions);
> - Increase the legitimacy of the assessment, and guard against bias from particular interest groups;
> - Assure quality of assessment outputs;
> - Design the outreach and communication activities; and
> - Help to raise funds for the assessment.[55]

In the LTMS a Project Steering Committee was set up to establish the leadership of the Department of Environmental Affairs and Tourism (as it was then). It met for the first time on 15 March 2006. The team, under

[55] Walter Reid, Neville Ash, Elena Bennett, Pushpam Kumar, Marcus Lee, Nicolas Lucas, Henk Simons, Valerie Thompson, Monika Zurek (2002) 'Millennium Ecosystem Assessment Methods'. © Millennium Assessment Secretariat. http://www.fao.org/gtos/tems/diversity/MA-methods.pdf.

the leadership of the DEAT, was at first made up of department officials, the Energy Research Centre (which was contracted to run the process) under the leadership of Harald Winkler, and myself representing the future facilitation team. It was our first time together, armed with a budget and a mandate. Our first task was to beef up the Steering Committee and to make sure that the Committee was properly constituted. This will differ from process to process, but we were concerned with two matters:

1. The Committee would have to represent the relevant and most important departments in government: given that in South Africa the issue of energy would be key, this meant that the Department of Minerals and Energy would need to co-manage, and Treasury, Science and Technology would also need to hold key roles. We did not want any inter-departmental problems.
2. Secondly we needed to ensure that the team was demographically representative. This is important for a number of reasons, not least of all to ensure respect from both the South African government in the team's make up and authority, and confidence from the players we would eventually involve. We also wanted to transfer capacity: the department included a number of government officials who were 'in development', so that capacity could be enhanced through their exposure to the process. This meant that throughout the process the question of internal capacity would be a priority. This is how the challenge was put at the time:

> In the field of Climate Change most South African expertise and knowledge resides with white male practitioners. In setting up the best team for the LTMS project, this equity imbalance would be perpetuated in the interests of top quality expertise. Hence the project must take other active steps to ensure that equity interests are promoted. This is primarily to be achieved through a process of 'partnered capacity building'. This partnered capacity building will in the first instance take place through a selection, at this stage and as and when appropriate, of key black players in government and elsewhere, to form part of the project team. Through direct mentoring these players will be included at all times in the unfolding of the project.

The truth of the matter was that we were all going to learn a great deal! Capacity development is always an important component:

> In the context of an assessment, capacity building is a continuous process aimed at strengthening or developing long-term relevant human resources, institutions, and organisational structures to carry out ecosystem assessments of relevance to decision makers and to act on the findings. Capacity building within an assessment has two objectives: to enhance the expertise of individual scientists to carry out ecosystem assessments and to enhance institutional expertise, particularly the science–policy interface, for effective adoption and use of the assessment findings.
>
> Assessments may also provide important capacity-building opportunities through improving research capacities of universities and other research and training institutions, establishing baseline data for further assessments in the future, fostering an appreciation for scientific knowledge on the part of decision makers, and establishing or strengthening regional networks of experts.
>
> Tangible ways to incorporate capacity building into the assessment implementation include developing a 'fellows program' for young scientists to partner with more senior or experienced scientists engaged in the assessment, providing training courses on Scenario (or other assessment) methodologies, developing training materials, and forming partnerships with other institutions to expand the reach of these activities and to address decision makers' needs to build their own capacity to use the assessment's findings.[56]

By the second meeting the steering process was fully structured. Here all the headline decisions on process and research direction were taken – this gave Harald and I confidence, as well as the occasional knuckle rapping, and although we were given the freedom to plan the way forward, the immense weight of responsibility on Harald's shoulders would be lightened by the support and ratification of the Committee.

We also dealt with important matters such as progress maps and budgets.

---

56 Ibid.

Annexure 8

Team members identified for SBT 1

Name	From	Sector/description
Alan Hirsch	Presidency	Government
Alan Munn	Sustainable Business Manager, Engen	Liquid fuels
Andre van der Bergh	BHP Billiton	Mining, coal
Andrew Borraine	SA Cities	Cities
Bill Rowlston	Department of Water and Forestry	Government
Bongi Gqasa	Department of Public Enterprises	Government
Chris Moseki	Department of Water and Forestry	Government
Dick Kruger	Anglo American	Mining
Dipolelo Elford	Western Cape	Provincial government
Dr A Paterson	Aluminium	Metals
Dr Johan Ledger	SESSA	
Dr Joseph Matjila	Kumba Resources (Iron & Steel)	Manufacturing
Dr Laurraine Lotter	Executive Director, CAIA	Chemical sector
Dr Trevor Chorn	Corporate Environmental Sepcialist, Engen	Liquid fuels
Dr John Scotcher	Forestry SA	Forestry, paper
Hassan Mohamed	Presidency	Government
Herman J. van der Walt	Group Air and Climate Change Adviser, Sasol	Liquid fuels, coal
Ian Langridge	EIUG	Heavy energy users
Imraan Patel	Department of Science and Technology	Government
Jeff Subramoney	Department of Minerals and Energy	Government
Joanne Yawitch	Department of Environmental Affairs and Tourism	Government
Judy Beaumont	Department of Environmental Affairs and Tourism	Government
Justice Mavhungu	Department of Public Enterprises	Government
Kadri Kevin Nassiep	CEF/SANERI	Regulator
Kelebogile Moroka	Department of Environmental Affairs and Tourism	Government
Leila Mohamed	SEA	NGO

Name	From	Sector/description
Linda Manyuchi	Department of Science and Technology	Government
Litha Mcwabeni	Department of Public Enterprises	Government
Liza Roussot	Treasury	Government
Lize Coetzee	Department of Trade and Industry	Government
Lwandle Mqadi	SACAN	NGO
Lwazikazi Tyani	Department of Minerals and Energy	Government
Lydia Greyling	Department of Foreign Affairs	Government
Mandy Rampharos	Eskom	Utility
MC Moseki	Director: Water Resource Planning Systems	Government
Mike Edwards	Forestry	Forestry
Neva Makgetla	Labour	Labour
Nic Opperman	Director: Natural Resources, AgriSA	Agriculture
Nwabisa Matoti	Business Unity South Africa	Business
Peter Lukey	Department of Environmental Affairs and Tourism	Government
Professor Robin Barnard	Agriculture	Agriculture
Richard Worthington	Earthlife	NGO
Rod Crompton	NERSA	Regulator
Russel Baloyi	SALGA	NGO
Sakkie van der Westhuizen	SAPPI	Paper
Sharlin Hemraj	Treasury	Government
Stan Jewaskiewitz	Envitech Solutions	Waste
Tony Frost	WWF	NGO
Tony Surridge	Department of Minerals and Energy	Government
Tshenge Demana	Department of Trade and Industry	Government
Tsietsi Mahema	Department of Environmental Affairs and Tourism	Government
Wendy Poulton	Eskom	Utility
Y Stan Pillay	Senior Divisional Manager: Sustainable Development, Anglo Coal	Mining, coal

Annexure 9

A list of 'fairy godmother' technologies

Defining categories

Indicator	Mitigation potential			Viability in the time horizon		
Measure	Large: 100Mt CO_2/yr	Medium: 10Mt CO_2/yr	Small: 1Mt CO_2/yr	Near: 2010	Medium: 2025	Beyond: beyond 2050
Code	L	M	S	N	M	F

New renewable energy (Mitigation = M–L)
New wind (M)
Urban/micro wind, distributed generation (N)
Ocean (wave, tidal, ocean current, thermal) (M–F)
Geothermal (F)
Biomass gasification (N)
Heat pumps (N)
SWH for cooling systems (N)
Energy density of biomass (N)
Second generation biomass (M)
Solar chimney (N)
Building-integrated PV (N)

Super Tech (Mitigation = S)
Nano-technology storage & ultra capacitors (M)
Superconductivity
Peltier effect (M)
Pizo (pressure) electric (M)
Wireless transmission (B)
Livestock methane (N)

New fossils (Mitigation = M–L)
Underground coal gasification (M)
Ultra supercritical coal (B)
Kalahari Gas: coal-bed methane (M)
Methane management e.g. coal (M)

Grid reform (Mitigation = M–L)
Net metering (N)
Vehicle to grid (M)
AC to DC transmission (F)
Energy mix (F)

EE (Mitigation = L, L, S)
Increased energy efficiency (N)
System efficiency (N)
LEDs (N)
Induction heating (N)

Tech policy (Mitigation = S–M)
Air traffic management (M)
1 Watt standby power (N)
Shipping

Hydrogen (Mitigation = L)
Hydrogen from non-carbon source (M)
Hydrogen economy (storage/ distribution/use) (M)
Fuel cells (e.g. hydrogen/ solid oxide) (M)

Transport (Mitigation = M–L)
Air to super-fast rail (N)
Regenerative braking on trains (N)

Tech combinations (Mitigation = M)
Integration of energy technologies e.g. PBMR with CTL (M)
System integration of RE technologies (N)

Replacement materials (Mitigation = M to L)
Light metals for automotive & aerospace industry (N–M)
Materials recovery and recycling (N)

New nuclear (Mitigation = L)
Breeder reactors (?)/fusion (B)
Fast neutron reactors (N)

Communications (Mitigation = L)
New communication technologies (N)
Virtual holidays (F)

Alt generation (Mitigation = M)
Micro generation (N)

Hydro (Mitigation = M–L)
Imported hydro (M)

Space (Mitigation = L)
Space-based 'ideas' (B)

Sinks (Mitigation = M)
Algae pond CO_2 sink (M)

Storage (Mitigation = M)
Storage (N–M)

Social policies (Mitigation = L)
Social policies (N–M)
Dietary change
Air rationing
Localisation

Annexure 10

Media Statement by Marthinus Van Schalkwyk, Minister of Environmental Affairs and Tourism, Cape Town, 28 July 2008

Government outlines Vision, Strategic Direction and Framework for Climate Policy

The South African Government launched its long-term mitigation Scenario (LTMS) process on Climate Change in 2006. Findings and policy recommendations based on the LTMS were presented to the Cabinet at its Lekgotla last week. This is the culmination of two and a half years of work that involved stakeholders from Government, business, civil society and labour.

During the Cabinet Lekgotla, Government discussed the policy implications of the LTMS in detail. In response, Government has outlined an ambitious vision and adopted a pro-active and scientifically and economically robust policy framework that will ensure we meet the challenges of Climate Change in decades to come. It has set the strategic direction for climate action in South Africa.

Government's vision and the implementation of this policy framework will be the best insurance policy current and future generations will have against the potentially devastating impacts of Climate Change. By adopting this strategic direction South Africa takes a leading position in the developing world and demonstrates it is ready to shoulder its fair share of responsibility as part of an effective global response. The worst impacts of Climate Change can be avoided if the rest of the world takes up the challenge in a similarly serious way, with developed countries taking the lead.

The international negotiations on strengthening the climate regime after 2012 gained significant momentum at the talks in Bali in December 2007. This process will conclude in Copenhagen at the end of 2009. South Africa's LTMS process also establishes parameters for our post-2012 negotiating positions.

Science tells us the climate challenge is urgent and Government has therefore formulated a comprehensive domestic response based on the best available science, Scenario building tools, rigorous analysis of energy and non-energy emissions, the consideration of a wide range of mitigation options and potentials, adaptation planning and economic models. This is indeed cutting-edge work.

Science also tells us that an increase in global average temperature above 2°C poses a danger to all of us, but in particular the poor. To avoid the worst impacts of Climate Change we need to limit the temperature increase to 2°C above pre-industrial levels. We are already approximately 0.7°C above pre-industrial levels.

The world faces a global climate emergency. It is now clear that only action by both developed and developing countries can prevent the climate crisis from deepening. While developed countries bear most of the responsibility for causing the problem to date, developing countries – including South Africa – must face up to our responsibility for the future. Whilst we have different historical responsibilities for emissions, we share a common responsibility for the future.

The technical work done in the LTMS makes it clear that without constraints our emissions might quadruple by 2050. This is, in the most literal sense, not sustainable: If we continue with business-as-usual, we will go out of business. The alternative is a very challenging Scenario – to make it our goal to achieve what is Required by Science of a developing country. According to the IPCC Fourth Assessment Report, avoiding dangerous Climate Change requires developed countries to reduce their emissions compared to 1990 levels by 80–95% by 2050, and by 25–40% by 2020. In developing countries, substantial deviations below business-as-usual baselines are required.

The implementation of a combination of the three LTMS strategic options – in other words those that can be achieved with known technologies and at a relatively affordable cost – can deliver a *substantial deviation from business-as-usual emission trajectories* in South Africa. By committing to and implementing this vision and policy framework, Government will make a meaningful contribution to the international effort. It is more ambitious and detailed than what many countries in the current negotiation process have put on the table.

Government's vision

Government has outlined its vision for climate policy in the following terms:

1. In designing our policy for the transition to a climate resilient and low-carbon economy and society, we will balance our mitigation and adaptation response.
2. Our climate response policy, built on six pillars, will be informed by what is 'Required by Science', namely to limit global temperature

increase to 2°C above pre-industrial levels. The six policy direction themes are:
Theme 1: Greenhouse gas emission reductions and limits;
Theme 2: Build on, strengthen and/or scale up current initiatives;
Theme 3: Implementing the 'Business Unusual' Call for Action;
Theme 4: Preparing for the future;
Theme 5: Vulnerability and Adaptation; and
Theme 6: Alignment, Coordination and Cooperation.

3. We will continue to pro-actively build the knowledge base and our capacity to adapt to the inevitable impacts of Climate Change, most importantly by enhancing early warning and disaster reduction systems and in the roll-out of basic services, water resource management, infrastructure planning, agriculture, biodiversity and in the health sector.
4. GHG emissions must peak, plateau and decline. This means it must stop growing at the latest by 2020–2025, stabilise for up to ten years and then decline in absolute terms.
5. Over the long term we will redefine our competitive advantage and structurally transform the economy by shifting from an energy-intensive to a climate-friendly path as part of a pro-growth, pro-development and pro-jobs strategy.
6. Implementing policy under six themes will lay the basis for measurable, reportable and verifiable domestic emission reduction and limitation outcomes.
7. Overall, this would constitute a fair and meaningful contribution to global efforts. We would demonstrate leadership in the multi-lateral system by committing to a substantial deviation from baseline, enabled by international funding and technology.

Mitigation strategy

With reference specifically to our mitigation strategy, Cabinet adopted the following approach:

1. The Start Now strategic option as outlined in the LTMS will be further implemented. This is based, amongst others, on accelerated energy efficiency and conservation across all sectors, including industry, commerce, transport and residential, inter alia through more stringent building standards.
2. We will invest in the Reach for the Goal strategic option by setting ambitious research and development targets focusing on carbon-friendly technologies, identifying new resources and affecting behavioral change.

3. Furthermore, regulatory mechanisms as set out in the Scale Up strategic option will be combined with economic instruments such as taxes and incentives under the Use the Market strategic option, with a view to:
   - Setting ambitious and mandatory (as distinct from voluntary) targets for energy efficiency and in other sub-national sectors. In the next few months each sector will be required to do work to enable it to decide on actions and targets in relation to this overall framework.
   - Based on the electricity-crisis response, government's energy efficiency policies and strategies will be continuously reviewed and amended to reflect more ambitious national targets aligned with the LTMS.
   - Increasing the price on carbon through an escalating CO_2 tax, or an alternative market mechanism.
   - Diversifying the energy mix away from coal whilst shifting to cleaner coal, by for example introducing more stringent thermal efficiency and emissions standards for coal-fired power stations.
   - Setting similar targets for electricity generated from both renewable and nuclear energy sources by the end of the next two decades.
   - Laying the basis for a net zero-carbon electricity sector in the long term.
   - Incentivising renewable energy through feed-in tariffs.
   - Exploring and developing carbon capture and storage (CCS) for coal-fired power stations and all coal-to-liquid (CTL) plants, and not approving new coal-fired power stations without carbon capture readiness.
   - Introducing industrial policy that favours sectors using less energy per unit of economic output and building domestic industries in these emerging sectors.
   - Setting ambitious and, where appropriate, mandatory national targets for the reduction of transport emissions, including through stringent and escalating fuel efficiency standards, facilitating passenger modal shifts towards public transport and the aggressive promotion of hybrids and electric vehicles.

Process going forward: 2009 to 2012

Cabinet has mandated a clear path for the future. Milestones will include a national summit in February next year, the conclusion of international negotiations at the end of 2009 and a final domestic policy to be adopted by the end of 2010 after international negotiations have been completed.

The process will culminate in the introduction of a legislative, regulatory and fiscal package to give effect to the strategic direction and policy from now up to 2012.

Annexure 11

LTMS Cabinet media release presentation

In order to promote clarity and awareness, the Department of Environmental Affairs and Tourism also released a presentation and further explanation. Entitled 'The Government's Vision, Strategic Direction, and Framework for Climate Policy' the document first set out the realities of the Climate Science and the international and national imperatives.

On impacts:

Water	Increased water availability in moist tropics and high latitudes Decreasing water availability and increasing drought in mid-latitudes and semi-arid low latitudes
	0.4 to 1.7 billion · 1.0 to 2.0 billion · 1.1 to 3.2 billion · Additional people with increased water stress
Ecosystems	Increasing amphibian extinction · About 20 to 30% species at increasingly high risk of extinction · Major extinctions around the globe
	Increased coral bleaching · Most corals bleached · Widespread coral mortality
	Increasing species range shifts and wildfire risk · Terrestrial biosphere tends towards a net carbon source, as ~15% · ~40 % of ecosystems affected
Food	Crop productivity · Low latitudes Decreases for some cereals · All cereals decrease
	Increases for some cereals Mid to high latitudes · Decreases in some regions
Coast	Increased damage from floods and storms
	About 30% loss of coastal wetlands
	Additional people at risk of coastal flooding each year · 0 to 3 million · 2 to 15 million
Health	Increasing burden from malnutrition, diarrhoeal, cardio-respiratory and infectious diseases
	Increased morbidity and mortality from heatwaves, floods and droughts
	Changed distribution of some disease vectors · Substantial burden on health services
Singular events	Local retreat of ice in Greenland and West Antarctic · Long term commitment to several metres of sea-level rise due to ice sheet loss · Leading to reconfiguration of coastlines world wide and inundation of low-lying areas
	Ecosystem changes due to weakening of the meridional overturning circulation

0 1 2 3 4 5°C

Global mean annual temperature change relative to 1980–1999 (°C)

In South Africa, there was already a great deal of data that Climate Change would heavily impact our country. This is especially so for rainfall and soil moisture.

The report goes on to view the imperative to mitigate rapidly, at international level:

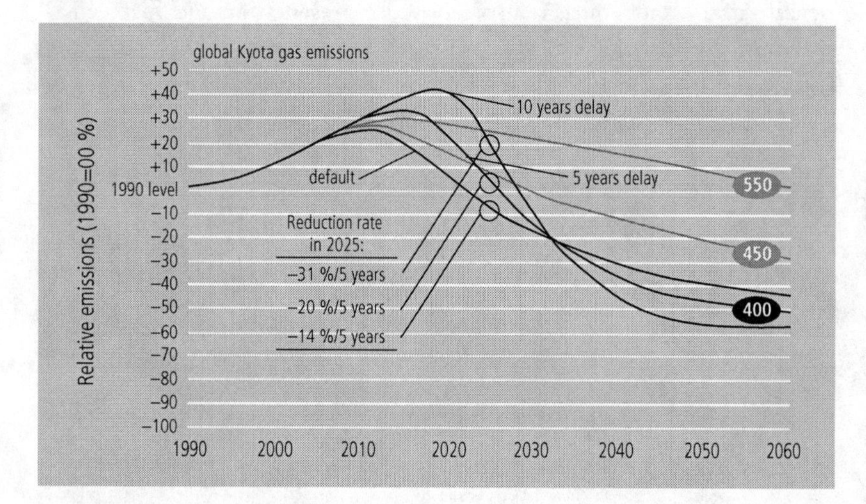

It then points out that this Required by Science approach includes a burden sharing discount:

The range of the difference between emissions in 1990 and emission allowances in 2020/2050 for various GHG concentration levels for Annex 1 and non-Annex 1 countries as a group

Scenario category	Region	2020	2050
A-450 ppm CO_2-eq	Annex 1	−25% to −40%	−80% to −95%
	Non-Annex 1	Substantial deviation from baseline in Latin America, Middle East and Centrally Planned Asia	Substantial deviation from baseline in all regions
B-550 ppm CO_2-eq	Annex 1	−10% to −30%	−40% to −90%
	Non-Annex 1	Deviation from baseline in Latin America and Middle East, East Asia	Deviation from baseline in most regions, especially in Latin America and Middle East
B-650 ppm CO_2-eq	Annex 1	0% to −25%	−30% to −80%
	Non-Annex 1	Baseline	Deviation from baseline in Latin America and Middle East, East Asia

It recognises the global shift to the low-carbon economy, and that South Africa will have to find opportunities in a carbon-constrained world, avoiding the risks and turning the country's potential comparative advantages into competitive advantages.

The four Scenarios of the LTMS study were now, correctly stated in a group, four strategic and cumulative Options:

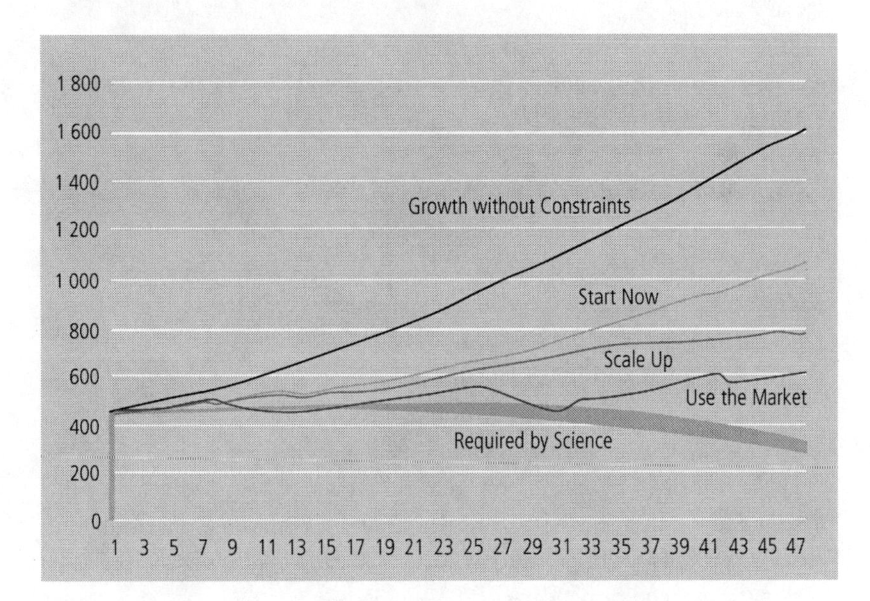

The economy wide modelling was restated in the following terms:

Start Now

- GDP impacts negative over the period – less than 1/10th of a per cent; Modelling does not fully account for savings from energy efficiency being spent elsewhere in the economy;
- Pattern of socio-economic impacts is confirmed – decreases in jobs for lower-skilled households;
- However, most households are better off due to lower energy prices.

Scale Up

- High growth effect due to higher levels of investment;
- GDP impact even more positive (from 1 to 1.3%) than under static model;
- Wage income increases for all skills groups (between 17% and 29%);

- Welfare improves for low-income groups, with a decline in welfare among richer households who derive most income from capital, not wages.

Use the Market

- Impact on GDP is mildly positive (0.73%) instead of the previous minus 2%;
- Price increases are overshadowed by higher investments;
- Income from employment increases for all household groups; and
- Differences in welfare effects are marginal.

The six broad policy directions

Theme 1: Greenhouse gas emission reductions and limits
Theme 2: Build on, strengthen and/or scale up current initiatives
Theme 3: Implementing the 'Business Unusual' call for action
Theme 4: Preparing for the future
Theme 5: Vulnerability and adaptation
Theme 6: Alignment, coordination and cooperation

were restated, and in Theme 1, Climate Change mitigation interventions should be informed by, and monitored and measured against the following 'plateau and decline' emission trajectory, with Greenhouse gas emissions ceasing to grow (start of plateau) in 2020–2025, and beginning to decline in absolute terms (end of plateau) in 2030–2035.

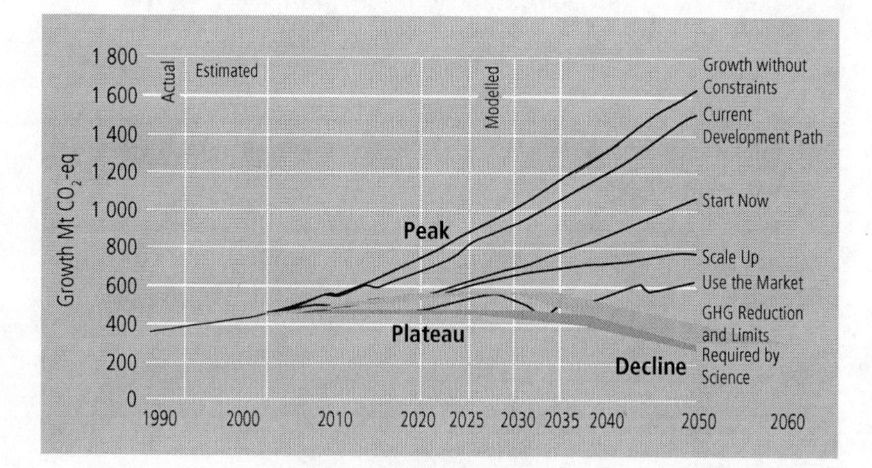

The remaining themes were expressed in the following terms:

Theme 2: Build on, strengthen and/or scale-up current initiatives

- Current energy efficiency and electricity demand-side management initiatives and interventions must be scaled-up and reinforced through available regulatory instruments and other appropriate mechanisms (*made mandatory*).
- Based on the electricity-crisis response, government's *energy efficiency* policies and strategies must be continuously reviewed and amended to reflect more *ambitious national targets* aligned with the LTMS.
- Treasury will study a *carbon tax* in the range modelled by the LTMS, starting at low levels soon and escalating to higher levels by 2018/2020, with sensitivity to higher and lower tax levels, and report to Cabinet on its findings.

Theme 3: Implementing the 'Business Unusual' call for action

- The renewable energy sector is identified as a key 'Business Unusual' growth target for renewable energy.
- The *transport sector* is identified as another key 'Business Unusual' growth sector and policies and measures are put in place to meet ambitious and mandatory national targets for the reduction of GHG emissions from this sector.
- In committing to national GHG emission limitation and reduction targets, govemment must promote the *transition to a low-carbon economy* and society and all policy and other decisions that may have an impact on South Africa's GHG emissions must take this commitment into regard.

Theme 4: Preparing for the future

- There is increased support for the new and ambitious *research and development targets* that are being set, especially in the field of carbon-friendly technologies – with the focus on the renewable energy and transport sectors.
- Formal and informal forms of *education and outreach* are used to encourage the behavioural changes required to support the efficient and effective implementation of the climate change response policy.

Theme 5: Vulnerability and adaptation

- South Africa continues to identify and describe its *vulnerabilities* to Climate Change.
- We describe and prioritise what *adaptation interventions* must be initiated, who should be driving these interventions and how implementation will be monitored.
- Affected government departments will ensure that Climate Change adaptation in their sectors are included as departmental *key performance areas.*

Theme 6: Alignment, coordination and cooperation

- The *roles and responsibilities* of all stakeholders, particularty the organs of state in all three spheres of government will be clearly defined and articulated.
- The structures required to ensure *alignment*, coordination and cooperation will be clearly defined and articulated.
- Climate change response policies and measures are *mainstreamed* within existing alignment. Coordination and cooperation structures.

The process going forward was set out:

Process going forward: 2009 to 2012

- National Climate Change Response Policy Development Summit (February 2009) (Adopt Framework).
- Sectoral policy development work (February–June 2009).
- Post-2012 negotiation positions (up to July 2009).
- UNFCCC post-2012 negotiations concluded (Copenhagen, December 2009).
- National policy updated for implementation of international commitments (March 2010).
- Green Paper published for public comment (April 2010).
- Final National Climate Change Response Policy published (end 2010).
- Policy translated into legislative, regulatory and fiscal package (from now up to 2012).

In three further passages, the government's vision for the road ahead was set out:

1. Transition to climate resilient and low-carbon economy and society – balance our mitigation and adaptation response;

2. Our climate response policy, built on six pillars, will be informed by what is Required by Science – to limit global temperature increase to 2°C above pre-industrial levels;
3. Continue to pro-actively build the knowledge base and our capacity to adapt to the inevitable impacts of Climate Change, most importantly by enhancing early warning and disaster reduction systems and in the roll-out of basic services, infrastructure planning, agriculture, biodiversity, water resource management and in the health sector;
4. GHG emissions must peak, plateau and decline – stop growing at the latest by 2020–2025, stabilise for up to ten years, then decline in absolute terms;
5. Long term: redefine our competitive advantage and structurally transform the economy by shifting from an energy-intensive to a climate-friendly path as part of a pro-growth, pro-development and pro-jobs strategy;
6. Implementing policy under the six themes will lay the basis for measurable, reportable and verifiable domestic emission reduction and limitation outcomes; and
7. This would constitute a fair and meaningful contribution to the global efforts, demonstrating leadership in the multi-lateral system by committing to a 'substantial deviation from baseline', enabled by international funding and technology.

On mitigation, our immediate task is to:

Start Now based on accelerated energy efficiency and conservation across all sectors (industry, commerce, transport, residential – including more stringent building standards); invest in Reach for the Goal by setting ambitious research and development targets focusing on carbon-friendly technologies, identifying new resources and affecting behavioural change; combine regulatory mechanisms under Scale Up; and economic instruments (taxes and incentives) under Use the Market with a view to:

1. Setting ambitious and mandatory (as distinct from voluntary) targets for energy efficiency and in other sub-national sectors. In the next few months each sector will be required to do work to enable it to decide on actions and targets in relation to this overall framework;
2. Based on the electricity-crisis response, government's energy efficiency policies and strategies must be continuously reviewed and amended to reflect more ambitious national targets aligned with the LTMS;
3. Increasing the price on carbon through an escalating CO_2 tax, or alternative market mechanism;

4. Diversifying the energy mix away from coal whilst shifting to cleaner coal, e.g. by introducing more stringent thermal efficiency and emissions standards for coal-fired power stations;
5. Setting similar targets for electricity generated from both renewable and nuclear energy sources by the end of the next two decades;
6. Laying the basis for a net zero-carbon electricity sector in the long term;
7. Incentivising renewable energy through feed-in tariffs;
8. Exploring and developing carbon capture and storage (CCS) for coal-fired power stations and all coal-to-liquid (CTL) plants, and not approving new coal-fired power stations without carbon capture readiness;
9. Introducing industrial policy that favours sectors using less energy per unit of economic output and building domestic industries in these emerging sectors; and
10. Setting ambitious and, where appropriate, mandatory national targets for the transport emissions, including through stringent and escalating fuel efficiency standards, facilitating passenger modal shifts towards public transport and the aggressive promotion of hybrids and electric vehicles.

Author biography

Stefan Raubenheimer is the Executive Director of SouthSouthNorth (SSN), and has worked in the field of Climate Change for ten years. He started his career as a human rights lawyer in the 1980s, during which period he trained as a mediator, arbitrator and facilitator. In the period before and after the South African elections in 1994, he became interested in the resolution of conflict and the facilitation of development and transformation processes. In this time he worked as a panellist for the Independent Mediation Service of South Africa, now Tokiso. In 1999, he attended the UN Climate Change Conference and, together with an international team, created SSN, an organisation that he has headed since then. SSN has focused on the developmental issues raised by the Climate Change challenge. A combination of his work in this new field and his work as a facilitator laid the seeds for the design and management of the LTMS, the subject of this book.